Food Packaging
Principles and Applications

NIPA® GENX ELECTRONIC RESOURCES & SOLUTIONS P. LTD.
New Delhi-110 034

Food Packaging
Principles and Applications

Kshitiz Kumar
Department of Food Processing Technology
A D Patel Institute of Technology
CVM University, V V Nagar
Gujarat, India

Pravin M Ganorkar
Department of Food Processing Technology
A.D. Patel Institute of Technology
CVM University, V V Nagar
Gujarat, India

Vijay S. Sharanagat
Department of Food Engineering
National Institute of Food Technology Entrepreneurship and Management
Kundli, Sonipat, Haryana, India

NIPA® GENX ELECTRONIC RESOURCES & SOLUTIONS P. LTD.
New Delhi-110 034

NIPA® GENX ELECTRONIC RESOURCES & SOLUTIONS P. LTD.

101,103, Vikas Surya Plaza, CU Block
L.S.C. Market, Pitam Pura, New Delhi-110 034
Ph : +91 11 27341616, 27341717, 27341718
E-mail: newindiapublishingagency@gmail.com
www: www.nipabooks.com

For customer assistance, please contact

Phone: + 91-11-27 34 17 17
Fax: + 91-11-27 34 16 16
E-Mail: feedbacks@nipabooks.com

ISBN: 978-93-95319-32-4

Composed and Designed by NIPA®.

Preface

Packaging is an essential part of food business and directly related to the food safety and shelf-life of food products. Food Packaging acts as an interface between consumer and product, makes the first impression on the consumer and is termed as "Signature of the packed food". Packaging material like paper, paper board, plastic, glass and tin are used in food packaging. Each of them has their own advantages and limitations. Selection of suitable packaging material depends upon the physiochemical properties of product to be packed and desired shelf life. Hence, the basic understanding of different food packaging materials and their properties become essential for the food business operators.

Food packaging is taught as specialized course in diploma, degree and postgraduate programmes of Food Technology/Food Processing Technology/ Food Engineering/Dairy Technology/Agricultural Process Engineering. The book 'Food Packaging: Principles and Applications' has been conceptualized to cover the syllabus of undergraduate course in these programmes. There was need to present the basic as well as novel topics of food packaging in simple way so that undergraduate student can comprehend them easily. The book has been written in simplified manner and the topics presented systematically to meet this requirement.

The book consists of 13 chapters covering different topics of food packaging. First few chapters of the book covers the principles of various forms of the food packaging material such as plastic, paper, metal and glass. The manufacturing process, properties and the application of each are also discussed. The packaging requirement of various groups of food products such as fruit, vegetables, milk, cereals, snack, and beverages is discussed in a separate chapter. Some special topics and essential learning in packaging such as 'shelf life of packaged food, aseptic packaging and recycling of packaging material' are also present. In order to synchronize students with advancement in food packaging the chapters on active packaging, intelligent packaging, modified atmosphere packaging, edible and biodegradable packaging has been added. The last two chapters of the book cover the laws on food packaging and lebelling to acquaint the students with the legal requirements of food packaging.

Editors acknowledge the effort of all the researches and teachers who have contributed in the book and are grateful to them for their hard work and enthusiasm in the book completion. We have tried to present the technical aspect in simple manner, however, feedback in any form to improve the quality of the book and its usefulness to the students are welcome.

Editors

Contents

1

Introduction to Food Packaging

Kshitiz Kumar

Department of Food Processing Technology
A D Patel Institute of Technology, CVM University, V V Nagar, Gujarat, India

1.1. Introduction

Food packaging involves enclosing the food product to ensure its safe delivery to end consumer in safe condition. Packaging is one of the integral components of the food industry. In order to maintain the basic nutritional and functional attributes of food material, they must be packaged safely. A faulty package can ruin all the efforts made to develop the food product. Selection of right packaging system and packaging material is essential to arrest the deterioration of food products from physical, chemical and biological factors. The developments in packaging systems and design have kept a rapid pace in the last fifty years. From merely wrapping the food in piece of thing whatever in hand, food packaging has today acquired a form of art and science. Food industry whether multinationals or local use different elements of food packaging to minimize the chance of adulteration and also to brand their products.

1.2. Developments in Food Packaging

The important historical developments in food packaging in the last 200 years include:

Year	Development
1800-1850	• Hermetically sealed glass jars by Nicolas Appert (1809) • Soldered tinplate by Peter Durand (1810)
1851-1900	• Paper bag-making machine by Francis Wolle (1852) • Corrugated materials for packaging by Albert L. Jones (1971) • Use of folding box (1884) • Crown cap for glass bottles (1892)

Year	Development
1901-1950	• Paraffin wax coated paper milk containers (1906) • Regenerated cellulose film (1912) • Cellophane (1927) • Canned beer (1935) • Polyethylene (PE) was produced (1939)
	• Aerosol can; Polyvinylidene chloride (PVdC) or Saran used as a moisture barrier resin (1946); • Dole Aseptic Process (Late 1940s)
1951-1960	• Retort pouch for heat-processed foods • Polypropylene is invented • Use of Polyethylene extensively in packaging, • Boil in bags for frozen food (1960) • Tear-off aluminum ends for composite juice cans (1960)
1960s	• High-density polyethylene milk bottles, Flexographic printing for foil (1961); • Paper-foil-plastic laminated containers called tetrahedron by 'Tetra Pak' (1962) • Shrink-wrapped corrugated trays for canned goods (1964) • Two-piece aluminum cans, Tin-Free-steel (TFS chromium) cans (1965) • Polyvinyl chloride (PVC) for food use (1967) • Stand-up pouches, Boil-in-bag packaging, Interlocking/stacking plastic containers, (1968); • Aluminum can recycling (1970)
1970s	• Blow-fill-seal technology (1972) • Stretch blow molding process(1974) • Retort pouches (1977)
1980s	• Controlled-atmosphere oxygen-permeable produce bag, (1981) • Multilayer (squeezable) plastic bottle for ketchup, Time/temperature indicator tags (1983) • Microwavable retort pouches (1986) • Holographic labels (1987)
1990s	• Three-dimensional label (1996) • Self-heating container. Oxygen-scavenging film (1999) • Polylactic acid (PLA) polymer made from renewable resources (2000)
2000s	• Radio frequency identification tags (2003); • Biodegradable shrink film (2006)

In the last twenty years, the change in consumer trends (demand of mild processed, convenient, extended shelf-life food), retail chain practices (long distance transport of food) and environmental concerns have driven the food packaging innovations. The development of active and intelligent packaging, biodegradable and green packaging, packaging materials and components from food by-products are the recent trends in which the research is driven.

1.3. Industry in India and World

Food packaging industry is the driving force in the overall growth of packaging industry. Even in growing economies like India, today people are ready to spend their money on packaged food. The increase in demand of packaged food in India is due to:

a. Rapid change in food preferences of consumers

b. Increase in income and rising disposable income

c. Increased urbanization and hectic lifestyle

d. Preference of fast food

e. Increased number of working women

f. Increased export of spices, grains, fruits and vegetables

g. Increase in popularity of the service industry where food either raw or cooked is delivered at door step is adding new dimension in the growth of food packaging industry.

The global food packaging market has a value of approximately US$ 345 billion in 2021. Considering the compound annual growth rate of 5.85%, the market value is estimated to reach a value of US$ 483 billion in 2028.

1.4. Basic Terminology

Packaging: The word 'packaging' has been defined in many ways each laying emphasis on different aspects. Lockhart (1997) has defined packaging as a socio-scientific discipline that operates in society to ensure the delivery of goods to the ultimate consumer in the best condition intended for their use. Packaging Institute International (now defunct) has defined packaging as the enclosure of products, items or packages in a wrapped pouch, bag, box, cup, tray, can, tube, bottle or other container form to perform one or more of the following functions: containment, protection, preservation, communication, utility and performance.

Package and packing: Package refers to the physical entity that contains the product. Packing refers to the action involving keeping the product in the package.

Packaging strategy: It refers to the plan which address all the aspects and activities involved in safe and cost effective delivery of packaged product to the end user. The various stakeholders of devising the packaging strategy include management, procurement, technical professionals, manufacturer marketing representatives and legal experts.

1.5. Levels of Packaging

Depending on how close the food product is with the packaging materials there are four levels of packaging:

1.5.1. Primary Packaging

The packaging materials which remains in direct contact with the food materials is called primary packaging. They are also referred to as 'retail' or 'consumer' packaging. It provides protection to food product and inform the consumer about product's details, uses and features. Examples: cans for beverage, bottles and pouches for jam, sauce, pasta box.

1.5.2. Secondary Packaging

It refers to the additional packing given to the product for protection and marketing. It may contain one or more primary package. It is the package used for distribution and handling of product. Example: A cartoon box containing large number of cans. The can is primary packaging while the cartoon box is the secondary packaging.

1.5.3. Tertiary Packaging

It refers to the packaging, which is used to provide protection to primary and secondary package during transport and storage. It is also called 'bulk' or 'transit' packaging. It also helps to move the inventory smoothly and ease their handling. This is generally not seen by end consumer. Example: Shrink-wrapped box on pallet, shipping box containing cartoon.

1.5.4. Quaternary Packaging

The large metal containers used to handle large numbers of tertiary package for long distance transport are referred to as quaternary packaging.

1.6. Functions of Packaging

The functions performed by packaging are multidisciplinary. Packaging performs the basic functions such as containment, protection and convenience of the product. In the last fifty years the functions of packaging has evolved with advancement in technology and changing consumer demands and expectations. Now a food package is also expected to play a active role in preservation of food product as well as communicating the consumer about the state of food material inside it, its history from filling stage to storage, transport and retail display. Package also helps in brand building and promotion of food product.

There are four basic functions of packaging: containment, protection, convenience, communication.

1.6.1. Containment

A package should be sufficiently large enough and of right shape to fit the product inside it. Other function of the package such as protection, preservation and communication can only be performed if the product is contained by the package. Without proper containment, product spillage and loss would be widespread. The spillage or leakage of the packed product due to inefficient fitting of the product in the container can cause substantial damage to the environment too. Hence, a properly packed food package contributes in protecting the environment too.

1.6.2. Protection

The packaging material protects the product packed inside it from the extremities of mechanical stress and environmental factors such as light, moisture, oxygen, and microorganism. This is regarded as the primary function of the packaging. The degree of protection provided by the packaging depends upon the mechanical and barrier properties of the packaging material. The protection afforded by the package is an essential part of the preservation process. Packaging helps to prevent the deterioration of food product from the chemical, enzymatic and microbial spoilage and subsequently preserve the food until it reaches the end user. This is only possible when food product remain protected inside the package. Therefore, protection and preservation function of packaging goes hand in hand. Packaging also protects or conserves much of the energy expended during the production and processing of the product. Packaging prevents food wastage and helps to ensures food security. It is estimated that extent of food wastage, if the food is packaged is less than 1 % while the wastage climbs to 10-20 % in the case of unpackaged food. The protection role of package does not end just by preventing the food spoilage and wastage rather it goes much beyond that. It protects consumer faith over the product and brand. It protects and controls the price of the material as well as other economical loss associated with wastage such as replacement, insurance and litigation. The motto of one of the leading packaging industry Tetra Pak is rightly defined, as '*a package should save more than it costs*'.

1.6.3. Convenience

It refers to the ease the packaging system provides to the consumer in handling, using and finally disposing the packed product. Today's consumer has some specific packaging demands as per their lifestyle such as ready to eat and ready to cook packaged foods, resealable package, sufficiently small packaged food to fit in their outdoor activities. For the packaged liquid food such as canned beverages, the consumer expects that opening of the seal be smooth and pouring (if required) be without any spillage. The uses of zip locking, microwavable packaging, resealable pouch, easy open aluminium, recyclable packaging can are some example of convenient function of packaging.

Designing of package size to fit the consumer requirement such as 250 ml bottled drinking water or 20 g butter slice is referred to as 'apportionment function of packaging' where the package size is cut to meet the consumer's required dimensions. The apportioned package has deeper penetration in different sections of the society.

The size and shape of the primary packaging should be such that it is convenient in their handling and transportation. Primary package should be unitized optimally in secondary package and the secondary package into tertiary package.

1.6.4. Communication

The consumer first interacts with the package before they experience the product packed inside it. The packaging provides all the necessary information to the consumer such as type of product, ingredients, quantity, nutritional information, manufacturer details, use-by date, usage method. Package acts as "silent salesman" and develops the brand value of the product. The specific design and shape of the package, the graphics, symbols, illustrations and the labels printed on it helps the consumer to instantly recognize the product. Today when consumer shops in supermarket, these attributes are very helpful in selecting the product and reduce the shopping time. The package has a visual impact on the consumer and hence a poorly packaged product is going to have no takers, even if the product inside it is of superior quality. The use of universal product code (UPC) on the package helps to scan the product and ensure quick dispersal of the product in supermarkets. Similarly, use of radiofrequency identification tags (RFID) helps to keep a track of the product during the supply chain.

1.7 Packaging and Society

The packaging has added a great value to the modern lifestyle. In modern societies especially in developing countries, the increase in number of women workforce has reduced the available time for household activities including cooking. Development of packaging systems, which ensures the availability of processed food in ready to use and ready to cook form, has helped to save the time. At the same time the disposal of waste packaging material after its use in the environment has adversely affected the ecosystem and has attracted lot of attention and criticism by different section of the society.

Following are the benefits of food packaging to the society:

i. Prevents spoilage and wastage of food products, maintains the nutritional profile of food and ensures the supply of safe food to the consumer thereby protecting their health.

ii. Extends the shelf life of the food.

iii. Helps to controls the price of the food product by reducing the wastage. The saving and profit to the industry is ensured by large scale production in packaged form which has deep penetration and acceptability in the market.

iv. Minimizes the adulteration in food.

v. Ensures the availability of food in hygienic and aesthetically presentable form.

vi. Provides the food in the form in which they can be easily opened, consumed and disposed.

vii. Helps in development of various processed foods which adds variety and offers alternatives to the consumers.

viii. Helps development of modern retail outlets, supermarkets where the packaged foods are sold.

1.8. Concern Related to Packaging

The use of large number of packaged food has also led to voluminous quantity of plastic, metal, paper and glass material to be disposed off. In absence of proper recycling strategy and facilities, these are dumped in land or water bodies leading to pollution. In many packaging system there are lot of free space in the packaging while in many cases excess packaging material are used to provide extra protection. These all leads to unwanted use of available resources. There are concerns about the misinformation and concealing the information on the package labels. The migration of certain components from the packaging material to food may be cause of contamination of the food product.

1.9. Packaging Design and Development

Design of packaging system for a food material requires coherent and synchronized work of different stakeholders such as consumer, management, market representative, technical professionals, raw material supplier, manufacturer, distributor and retailer. A proper project planning for the packaging should start at the product concept stage itself. The design approach should be based on consumer, market and technical research. The challenge in packaging design is to achieve packaging optimization by striking a balance between cost and quality. The economics of design should incorporate every cost involved including disposal and recycle after product use. The packaging strategy, design and development should consider following requirements: a) Product requirements b) Distribution Requirement c) Machinery and Production Requirement d) Consumer Requirement e) Market Requirement f) Environmental Requirement

1.9.1. Product Requirements

The packaging design must consider the product characteristics in terms of its physical state, mechanical strength, chemical properties and microbiological status. The product design of powdered food product would be different from food, which are in emulsion form. Fragile food such as baked goods must have sturdy secondary packaging to protect them during transport and handling. Package containing food materials prone to oxidation must keep oxygen and light away. The package must protect and preserve the food till its shelf life.

1.9.2. Distribution Requirement

The time from which the food is filled inside the package till it reaches the customer constitute the distribution chain. During the distribution channel, the package interacts with various environmental factors, which can be broadly, be categorized into three groups: climatic, physical and biological.

1.9.3. Climatic Environment

It encompasses all the climatic factors such as air (gases such as O_2), moisture, light, temperature, humidity, pressure through which the food interacts in the distribution channel. The permeability of the package has to designed such that it allows only the required amount of gases and water vapour to pass through them. Protection against light (visible, infra-red and ultra violet rays) which catalyses several biochemical reaction has to be designed through package. Similarly, package are exposed to high and low temperature, relative humidity variations as they move from processing unit to end user. The change in altitude causes pressure variation and package has to endure such extremities too.

1.9.4. Physical Environment

The mechanical hazards such as compression, vibration, shock, abrasion, puncture and tearing which causes physical damage to the packaged products during transport and storage constitutes the physical environment. Loading, unloading, stacking of the packaged products in the transport vehicle, warehouses and retail outlet exposes them to various degree of compression, impact and shear force. When these forces exceed the yield point of packaging material, tear, puncture and failure will take place. Hence, packaging material of adequate tensile strength, stiffness, tear resistance and puncture resistance has to be selected. More the bulk storage point in the distribution chain, greater will be manual handling and risk of damage due to physical environmental factors. The design of packaging must consider the physical environment the package would encounter and accordingly provide the required mechanical strength to the packaging material.

1.9.5. Biological Environment

The biological environment through which the package interacts are microorganism, insect, pest, rodents, mites and human beings. The interaction with human is covered under consumer requirements. Microorganism such as bacteria, fungi, yeasts and viruses are ubiquitous and hence any loss in integrity of the package will result in food material being invaded by them.

1.9.6. Machinery and Production Requirement

Modern food packaging systems are integrated with the product processing where packaging machinery are located at the end of the production line. The packaging design has to also look into machinery used for forming the packaging material as well as all other integrated process such as filling, sealing, seaming, labeling, stretch wrapping and palletisation. For example, in form-fill-seal machine, which is widely used in food packaging, the roll width, core diameter, heat sealing properties, thickness, frictional, anti-static and anti-blocking properties of the packaging film has to be specified precisely for the smooth production process. Any change in the material of packaging will affect the machine performance. Similarly any change in can size or shape used in packaging will require suitable changes to be made in palletiser unit, conveyor guide rail system, filler and labeller unit, closure unit and casing unit.

1.9.7. Consumer Requirement

Consumer need and requirement is major force for new innovations in packaging systems. Consumer requirement and expectations keep changing with time and so does the packaging technology. Today's consumer demand may be contradictory in nature, for example, at one hand we wish minimally processed food, natural products with minimum processing while at the other hand we long for extended shelf life products. The packaging offers some degree of solutions satisfying both the requirements. Consumer need in terms of quality of the product in terms of retention of flavor, colour, texture, nutritional value and freshness of the products decides the packaging material and packaging system for any product. Convenience of the end user while handling the packaged product such as ease of access, opening, resealability, microwaveability and disposal the package plays a role in deciding the size, closure and material of the package.

1.9.8. Market Requirement

The emergence of modern retail outlets and supermarkets plays a big role in design of the package. The presentation of the product at retail outlet shelf and the marketing information to be displayed such as free extra product, price reduction, new improved version of the products plays the role in overall design of the package. Packaging helps in building the brand of the products. Package design has to be consistent with the current market drive. The market also decides the size of the package and manufacturer have to bring the product in various pack size as per market demand. For example, one of the reason for the introduction of snack food in small pack is the buying capacity in small town and villages.

1.9.9. Environmental Requirement

Increased environmental pollution due to careless disposal of the packaging material has put pressure on the packaging industry to adopt suitable strategies to reduce its ill effect on the environment. Packaging innovations such as biodegradable packaging, edible packaging, reduction in weight of packaging materials and strategies such as reuse, recycle are being adopted for sustainable packaging.

1.10. Trends in Food Packaging

i. Development of recyclable packaging material with their origin from sustainably managed forest.

ii. Minimize the transportation cost by looking for lightweight packaging material.

iii. To roll out packaging design and structure so that product is identified as unique and well differentiated on the shelf.

iv. To develop packaging design to engage consumers on personal level.

v. To develop active, intelligent and smart packaging systems.

References

Brown, W.E. (1992). Plastics in Food Packaging: Properties, Design and Fabrication. New York: Marcel Dekker.

Fellows, P. (2000). Freezing. In: Food Processing Technology Principles and Practice Second Edition (pp. 418-439). CRC Press, Boca Raton, Florida, USA.

Gupta, R.K., Dudeja, P. (2016). Food packaging. In: Dudeja, P., Gupta, R.K., & Minhas, A. S. (Eds.). Food safety in the 21st century: Public health perspective (pp. 547-553). Academic Press. https://doi.org/10.1016/B978-0-12-801773-9.00046-7.

Han, J. W., Ruiz-Garcia, L., Qian, J. P., & Yang, X. T. (2018). Food packaging: A comprehensive review and future trends. Comprehensive Reviews in Food Science and Food Safety, 17(4), 860-877.

Hanlon, J.F., Kelsey, R.J., Forcinio, H.E. (1998). Handbook of Package Engineering, 3rd edn. CRC Press, Boca Raton, Florida, USA.

Lalpuria, M., Anantheswaran, R., & Floros, J. (2012). Packaging technologies and their role in food safety. In Microbial Decontamination in the Food Industry (pp. 701-745). Woodhead Publishing.

Lange, J, Wyser, Y. (2003). Recent innovations in barrier technologies for plastic packaging: A review. Packaging Technology and Science 16: 149–153.

Lund, P.R., McCaul, J.P. (2009). Nitrile polymers. In: Yam, K.L. (Ed.), The Wiley Encyclopedia of Packaging Technology, 3rd edn.,New York: John Wiley & Sons, pp. 819–822.

Majid, I., Nayik, G.A., Dar, S.M., & Nanda, V. (2018). Novel food packaging technologies: Innovations and future prospective. Journal of the Saudi Society of Agricultural Sciences, 17(4), 454-462.

Mauer, L. J., Ozen, B.F. (2004). Food Packaging. In: Smith, J.S. and Hui, Y.H. (Ed.), Food Processing Principles and Applications (pp. 101-132). Blackwell Publishing, 2121 State Avenue, Ames, Iowa 50014, USA

Morris, S.A. (2011). Food and Package Engineering. Chichester, England: Wiley-Blackwell.

Robertson, G.L. (Ed.) (2010). Food Packaging and Shelf Life: A Practical Guide. CRC Press, Boca Raton, Florida, USA.

Robertson, G.L. (Ed.) (2013). Food Packaging Principles and Practice Third Edition. CRC Press, Boca Raton, Florida, USA.

2

Plastic Polymers in Food Packaging

Pravin M. Ganorkar

Department of Food Processing Technology, A. D. Patel Institute of Technology CVM University, New Vallabh Vidynagar, Gujarat, India

2.1. Introduction

The word 'plastic' is derived from the Greek word 'plastikos', which means easily shaped or molded. Plastics is defined as a random set of materials based on synthetic or modified natural polymers which can be formed by thermal and pressure treatment. It comprises of wide range of macromolecular organic compounds obtained either from alteration of natural macromolecules or polymerization of molecules of lower molecular weight.

The use of plastics as food packaging materials has increased its usage horizon dramatically in last two decades due to its remarkable properties like light weight, chemical inertness, diversity in shape and size, microbial attack resistance and importantly being economical. Food and beverages category lead the worldwide plastic packaging market in 2020 with 51% share followed by personal and household care products and pharmaceuticals. One time use plastic packages are witnessing substantial growth in food packaging due to its convenience. The success of convenience food and snack food like Indian namkeen products like sev, bhujiya, potato or banana chips etc. is due to affordable plastic packaging. Additionally, e-commerce business along with home deliveries of fresh fruits & vegetables, processed/cooked food is boosting the use of plastics as food packaging.

Plastics are suitably selected for definite technical use considering the particular requirement in storage, packing and supply chain. Moreover, plastics are chosen for specific product use, convenience and marketing purpose. Plastics are used in food packaging as flexible pouches, rigid containers, bulk product drum, lids, caps, trays, labels, tagger, plastic laminated paperboard, laminated aluminum foil, laminated pouches, flexible tubes, coating resins, strips, stretch & shrink wrapping, pallets, crates and adhesives. These plastics can be shaped, decorated or printed in several ways based on the type of packaging concerned. They can platter wide range of appearance and performance in the food packaging application, due to inherent properties of individual grade or combined effect of different grades with other packaging materials like paper, paperboard and metal. Plastic are used in single plastic grade and combined plastic grade. Single plastic grade consists of single macromolecular organic compound only while combined plastic grade consist of macromolecular organic compound combined with paper, metal or other macromolecular organic compound. Single plastic grade have advantages as well as certain limitations in the application. They are effective against a particular packaging requirement but cannot offer a wide range of packaging functionality. Combined plastic grades in the form of laminate provide excellent performance. Plastic films are combined by lamination, co-extrusion or coating to achieve the desirable properties which individual plastic grade cannot provide alone.

No doubt, on other side of coin, plastic possesses potential threat to environmental pollution considering waste disposal. In the present chapter, the fundamental plastic related terminologies and various plastics grade properties along with the major applications in food packaging has been discussed.

2.2. Historical Developments in Plastics

Charles Goodyear invented the process of vulcanization (rubber – sulphur reaction) in 1839. He obtained the patent for manufacturing hard rubber, commonly known as ebonite and vulcanite. It was considered as first theromoset plastic with discrete modification of natural rubber. Alexander Parkes introduced the first organic compounds plastic as “Parkesine” in 1862 as a substitute for ivory and horn at International Exhibition, London. It was manufactured by treating cotton waste with sulphuric acid and nitric acid mixture, and then further addition of camphor and castor oil. The resultant substance was used for making attractive articles. Further, John Wesley Hyatt manufactured the material, trademarked as ‘Celluloid’ and patented in 1870. Celluloid achieved notable commercial usage at early ages after introduction. Then progressively other plastics grades were introduced. Notably, Dr. Leo Hendrik Baekeland came up with the first synthetic plastic traded as ‘Bakelite’, the grade of phenol formaldehyde resins in 1907. German chemist Hermann Staudinger hypothesized that plastics are giant molecules in 1920 and proposed linear structure of polystyrene and natural rubber in 1924. This triggered the major breakthroughs in plastic chemistry. Staudinger was awarded the Noble prize for chemistry in 1953 towards the appreciation of his efforts and establishing polymer science. Today’s major plastics (polyvinyl chloride, polystyrene and polyethylene) were developed during 1930–1940. A number of plastics grades were innovated during World War II. Polyethylene was firstly manufactured in England in 1933. The company DuPont, introduced Nylon as synthetic silk in 1939, which was used by U.S. military for parachutes and ropes. Moreover, The Dow Chemist accidently discovered expanded polystyrene (EPS) in 1941, which later on was used as shock absorber and thermal insulator. The commercial use of polypropylene, one of the most used polymer, started in 1954 and it became a very useful polymer owing to diversity and adaptability. In the present era, plastics are renowned for strength, flexibility, designs and durability in various sectors. Specific purpose plastics are being discovered in the recent years than general purpose plastics. Present society is more relying on the plastics than other materials. The major user of plastics packaging materials is food industry. Efficient and cost effective plastic materials are paving the way substantially in the food packaging.

2.3. Basic Terminologies

i. Monomer: The small basic building unit of plastic structure

ii Polymer: The Greek word 'meros' means parts. Poly means many. Many parts joined together.

iii. Homo-polymer: Same repeating monomer (building block units) joined together.

iv. Hetero-polymer: Different two or multiple monomer unit throughout their molecules

v. Co-polymer: Two different monomers units are joined together

vi. Linear polymer: Polymer chain expand in one dimension

vii. Crosslink polymer: Polymer chains are joined together by relatively short cross links

viii. Syndiotactic polymer chain: The side groups of the polymer chain are arranged alternatively.

```
    X   X   X   X
    |   |   |   |
A-A-A-A-A-A-A-A-A-A
      |   |   |
      X   X   X
```

ix. Isotactic polymer chain: The side groups are located on the same side of polymer chain.

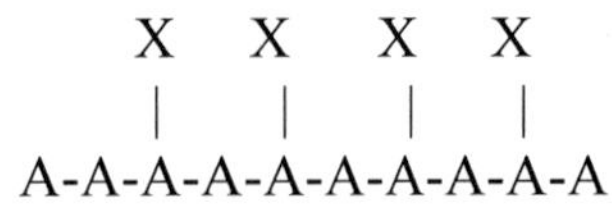

x. Atactic polymer chain: Random arrangement of side groups on the polymer chain

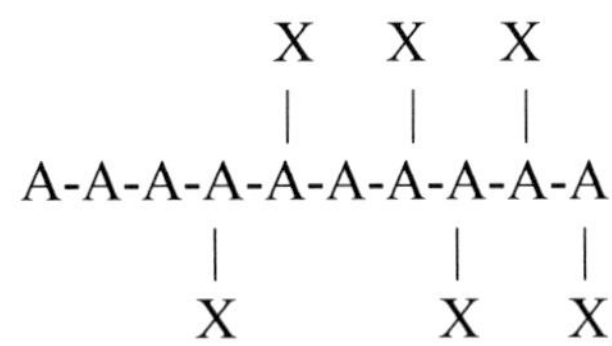

xi. Amorphous polymer: Random arrangement of molecular structure or unorganized molecular structure pattern, mainly constituting of atactic polymer chain

xii. Crystalline polymer: Ordered arrangement of molecular structure or organized pattern of molecular structure, mainly constituting syndiotactic or isotactic polymer chain

Difference between amorphous and crystalline polymer

Amorphous polymer	**Crystalline polymer**
• Polymer does not have an organized pattern in its molecular structure	• Polymer have an organized pattern in its molecular structure
• Mainly constitutes of atactic polymer chains	• Mainly constitutes of syndiotactic or isotactic polymer chains
• Weak Structure due to weak forces of attraction between molecules	• Compact structure due to strong forces of attraction between molecules
• Less Density	• High Density compared with amorphous polymer
• Chemical resistance is low	• Chemical resistance is relatively higher
• Transparent if native polymer	• Translucent to opaque
• Generally soft and less resistant to solvent penetration	• Generally rigid and relatively resistant to solvent penetration

xiii. Polymerization: The process of joining together a sequence of monomers under suitable temperature and pressure is referred as polymerization.

iv. Degree of polymerization: The average number of repeating units in a single molecule of a polymer is known as degree of polymerization (DP).

Molecular weight of polymer = Degree of polymerization X *Molecular weight of monomer*

As the DP increases, substance becomes greasy, wax and then above 1000 DP, it may solidify. The DP may go upto 1,00,000 or more.

2.4 Classification of Plastics

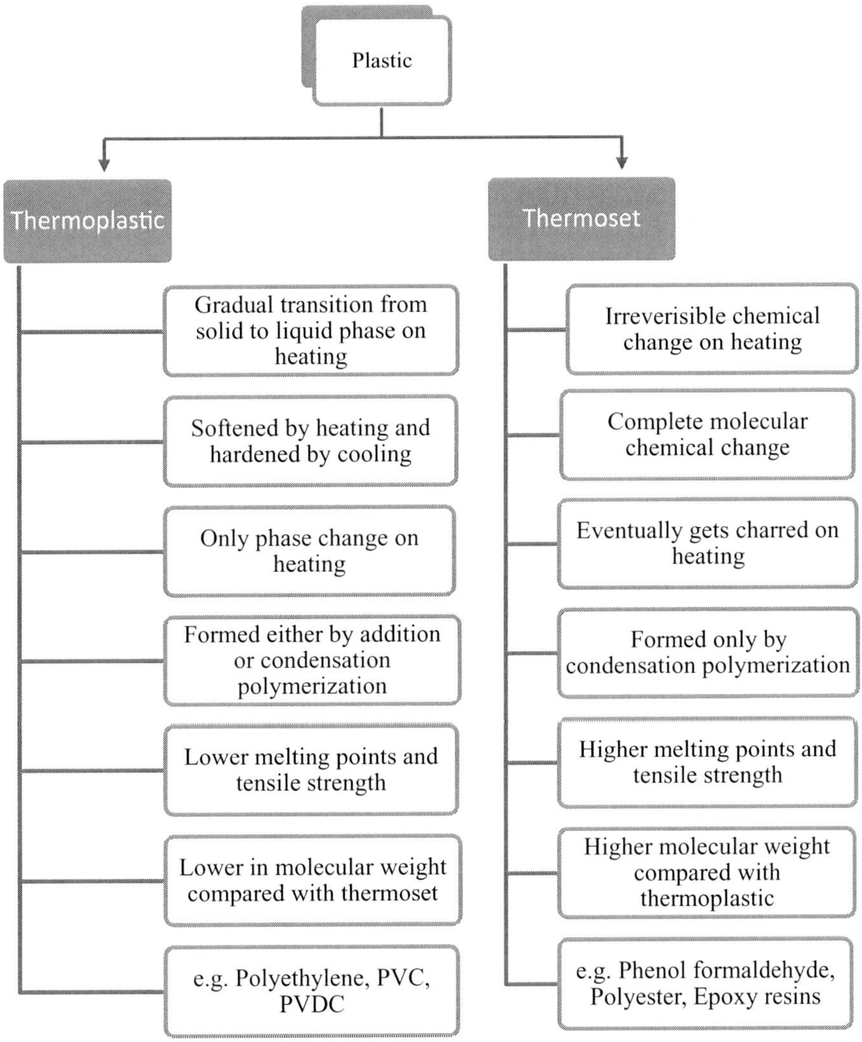

Plastics can be classified as thermoplastic and thermoset. Thermoplastic is comprised of long linear polymer joined together by weak van der Waals forces. Thermoplastic are manufactured either by addition polymerization or condensation polymerization. If they are heated, at first they become soft, and then further heating converts in liquid phase. After cooling, they can be re-solidify or hardened. Remolding is possible in thermoplastic. They are soluble in organic solvents. Usually they are soft, weak and less brittle. Thermoplastic are processed by injection, extrusion, blow, rotational moulding.

On other hand, thermoset is comprised of 3D network structure, more amorphous, cross-linked and constructed with strong covalent bonds. They are manufactured only by condensation polymerization. If thermoset are heated, they initially exhibit thermal resistance, but further heating causes irreversible chemical change in their molecular structure. Eventually they get charred or burned. Black ash may be visible in most of complete burning of thermoset. Their remolding is not possible. They are hard, strong and comparatively more brittle. They are insoluble in organic solvents. Thermoset are processed by compression and reaction injection molding. Generally thermoplastic are expensive than thermoset, as the manufacturing of thermoplastic is more energy intensive in terms of higher temperature and pressure. Thermoset have higher thermal stability once formed, compare to thermoplastic. Moreover, thermoset are relatively more chemical inert than thermoplastic.

Both categories of plastic are widely used as packaging material in food industries at different levels primary, secondary, tertiary and quaternary. Based on the properties of individual plastic grade, individual or in combination finds their suitability as food packaging materials. Further, the different plastic grades used in food packaging are discussed with respect to properties, application and limitations.

2.4.1. Polyethylene

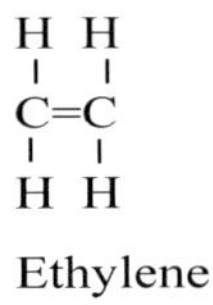

Ethylene

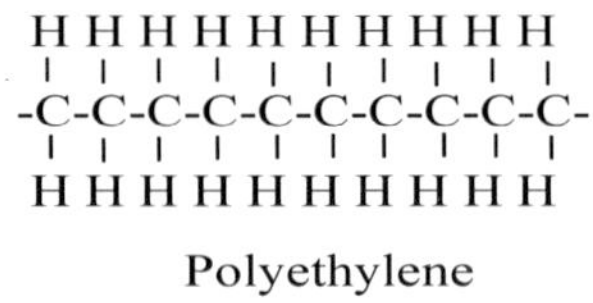

Polyethylene

Polyethylene is one of the largest and common plastic grades used in food packaging. It is utilized mainly for flexible food packaging. It comes under polyolefins category. It consists of ethylene monomers as structural building block and are categorized as homo-polymer. The free radical chain polymerization of ethylene (C_2H_4) is carried out at high temperature and pressure, in presence of metal alkali catalyst (Ziegler – Natta catalyst). Temperature may vary from 100 to 350 ^{0}C and pressure between 1000 to 3000 atm. The process may result atactic or isotactic polymer. Atactic polyethylene is more flexible and less tough due to less dense structure owing to branched structure. Isotactic polyethylene is more tough and highly dense due to dense structure owing to more linear structure. Therefore, isotactic polyethylene is more crystalline than atactic polyethylene and vice versa for amorphous nature.

Polyethylene can be graded on the basis of density in to three different categories.

Grade	Density
Low density polyethylene (LDPE)	915 – 925 kg/m^3
Medium density polyethylene (MDPE)	926 – 939 kg/m^3
High density polyethylene (HDPE)	$\geq$ 940 kg/m^3

However, at commercial level MDPE grade is not referred. It is merged with LDPE. Therefore, two grades of polyethylene i.e. LDPE (915 – 939 kg/m^3) and HDPE ($\geq$ 940 kg/m^3) are commonly used.

2.4.2. Low Density Polyethylene (LDPE)

LDPE is manufactured by polymerization of ethylene at high pressure. It consists of long and short chains due to some extent of degradation occurring at higher pressure. The branch chains leads to lose packing of the main polymer chains which finally results in low density polyethylene. Molecular weight and crystalline nature of LDPE eventually becomes lower. Moreover, this results in the lower softening point of LDPE (85–87 ^{0}C). This makes LDPE most appropriate choice for sealing layer in food packaging laminates. It melts easily and on cooling gives tight seals. It can be sealed even in presence of liquids. By virtue of this, LDPE is preferred as sealing layer. It has good combination of toughness, tensile, bursting, impact and tear strength. It also has excellent acid or alkali resistance (chemical resistance). But it has poor resistance to oils and greases. So, it has restricted use in edible oil flexible pouch packaging. However, high molecular weight grades of LDPE can be explored for fat rich food products. LDPE has a good barrier to water vapor transmission. The WVTR (water vapor transmission rate) for 100 micron thick LDPE film is 16-23 g/m^2/24 h at 38 ^{0}C and 90% RH. But, GTR (Gas transmission rate) is 7000-8000 cc/m^2/24 h at 25 ^{0}C and atmospheric pressure. Therefore, alone it is not suitable for foods prone to oxidation or foods which are highly hygroscopic. The printability of LDPE can be improved by treating surface with corona-arc discharge or flame techniques.

LDPE is also used as a rigid packaging material. Though it is rigid, its flexibility enables the contents to dispense after squeezing. Therefore, rigid form of LDPE is mainly used in collapsible tube; sprout caps varieties and other dispensers.

Different grades of LDPE are now available. Most of them differs in properties arising from varied degree of short chain branching, long chain branching, crystallinity, average molecular weight, presence of other co-monomer units. One of them most widely used in food packaging is linear low density polyethylene (LLDPE). Structurally, it contains multiple short chains. The co-polymerization of ethylene with a smaller amount higher alkenes (propene or hexane) results in LLDPE. Commonly, it has a density of around 920 kg/m^3. Short chains branching prevent the ability of polymer to crystallize and so it has similar density to LDPE. The branched structure provides toughness; on the other hand linear structure provides strength. It has better heat-seal strength but requires higher sealing temperature than LDPE. LLDPE performs better at high or low temperature compared to LDPE. It has also better chemical resistance, higher surface gloss, higher strength, puncture resistance and environmental crack resistance. Therefore, LLDPE is preferred in the packaging of frozen foods. Due to improved properties, LLDPE is leading the way in food packaging applications after LDPE and HDPE.

2.4.3. High Density Polyethylene (HDPE)

HDPE is manufactured by polymerization of ethylene at lower pressure and temperature compared with LDPE. It has more linear structure and crystalline than LDPE. It has white, translucent appearance. HDPE films are stiffer and harder than LDPE. It has up to 90% crystallinity, whereas LDPE exhibits crystallinities as low as 50%. The softening point is around 120 ^{0}C. Heat sealing requires high impulse heating compared with LDPE. Tensile strength and bursting strengths are higher but impact and tear strength are lower compared to LDPE. The chemical resistance of HDPE is superior and specifically has higher resistance to oils and greases. The WVTR for 100 micron thick HDPE film is 5-11 g/m^2/24 h at 38 ^{0}C and 90% RH. GTR (Gas transmission rate) is 1500-2000 cc/m^2/24 h at 25 ^{0}C and atmospheric pressure. Therefore, it offers better moisture protection and decreased gas permeability compared to LDPE. Due to better oil/fat resistance and barrier property, HDPE is commonly used for flexible edible oil pouches. It can also be used in rigid packaging form such as bottles. But, other plastic grades like Polyvinyl chloride and polyethylene terephthalate are dominating in this form due to their better barrier properties than HDPE. It is mainly used as a basic layer to make economical affordable material, still able to give stiffness to the packaging material.

The other popular grade of HDPE is high molecular high density polyethylene (HMHDPE). It has superior mechanical strength and barrier protection. While folding, it makes a characteristics sound compared with HDPE. It has good stiffness, high initial tear and propagating resistance. As it has improved barrier property to air and moisture, it is majorly used for the packaging of dehydrated or dried food products. HMHDPE can be easily handled by packaging machine due to high density compare with LDPE and HDPE.

The other special grade of polyethylene is irradiated polyethylene. It is manufactured by passing LDPE film through high energy β-rays emitted by electron beam accelerator. It gets converted to more cross-linked chain films, which gives higher strength in terms of stretch resistance and shrink tension due to hydrogen evolution and crystalline nature reduction. The film barrier property reduces at some extent, but makes it appropriate for shrink film. Irradiated polyethylene can shrink by passing through a hot air tunnel at 220 ^{0}C.

2.4.4. Polypropylene (PP)

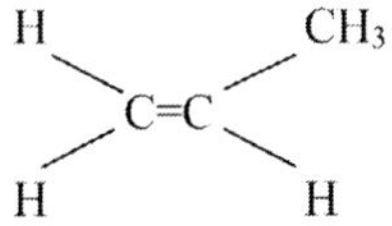

Propylene

The polymerization of propylene at 60 ^{0}C and about 100 atm pressure in presence of metal catalyst gives polypropylene. It is a linear polymer having higher molecular weight. Atactic PP (methyl groups are randomly placed) is amorphous and rubbery material which fetches a little value. Isotactic PP (methyl groups are placed on one plane) is stiff, rigid and more crystalline. It has lower density (900 kg/m^3). It possesses low water vapor transmission rate (7-9 g/m^2/24 h) and moderate gas transmission rate (2000-3000 cc/m^2/ 24 h). It has good resistance to oils and greases and also to abrasion. It has higher tensile strength, melting point and softening point (140-150 ^{0}C). Its high temperature stability makes it suitable for packaged processed food products where high processing temperature upto 130 ^{0}C is involved as in case of retorted pouches. Non oriented PP film is referred as cast PP film. The cost of cast PP film is much lower than oriented PP film. It is not recommended for use with heavy or dense products as its puncture resistance is less. PP film can be used as strapping tape and stretched filaments woven sack. PP woven sack are used in the packaging of grains and vegetables where gas and moisture in and out (breathability) is required, at the same time requires to prevent organic matter degradation, inertness and light weight. Cast PP can't be used at lower temperature as it becomes brittle below freezing temperature. This limitation can be overcome by addition of some amount of ethylene at the time of manufacturing. Printing can be carried out on PP film owing to high gloss and clarity.

The oriented grade of PP is referred as OPP film. Moreover, Bi-axially oriented PP (BOPP) film finds more application in the food packaging. The PP film is oriented in both direction (machine direction and transverse direction) while manufacturing. This orientation imparts high mechanical strength and excellent clarity compared to non oriented or cast PP film. The type and degree of orientation imparts wide range of properties. Bi-axial orientation substantially improves moisture barrier property. WVTR of 100 micron thick BOPP film is about 4 g/m^2/24 h at 38^{0}C and 90% RH difference. On other side, GTR is slightly reduced to about 2000-2500 cc/m^2/24 h at 25^{0}C. Outstanding resistance to water vapor transmission makes this film suitable for packaging of biscuits, cookies and other bakery and confectionary products. It has excellent tensile strength and tear strength. However, after initiation of tear, tear resistance is low. So, it is used in joint tapes. The limitation of the BOPP film is heat sealing. It can't be heat sealed on its own. This limitation can be overcome either by PVdC (Polyvinilidene chloride) coating as the same provides an excellent gas barrier or by co-extruding the film with layers of lower melting point polymer grade like LDPE. PVdC coated BOPP film can be employed where there is requirement of an excellent gas barrier. Co-extruded BOPP-LDPE film can be employed for low gas barrier applications. Very low density BOPP is especially used for wrapping confectionery instead of glassine paper.

2.4.5. Polyvinyl Chloride (PVC)

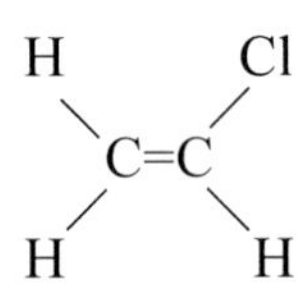

Vinyl Chloride

PVC is second most widely used polymer after polyethylene. It is commonly referred as vinyl. Vinyl compounds are olefins in which one hydrogen atom of ethylene is replaced. When two hydrogen atoms on the one carbon are replaced, it is referred as vinylidene. PVC is substituted olefin manufactured by polymerization of vinyl chloride monomers. First, ethylene dichloride is formed by addition of chlorine to ethylene and then it is dechlorinated or cracked to give vinyl chloride monomer. Mostly atactic form of PVC gets produced from polymerization. PVC film with wide properties can be obtained. Two main forms of PVC in food packaging are used i.e. rigid and plasticized form.

Rigid form of PVC is un-plasticized and used for trays, containers or bottle manufacturing. It degrades and discolor at temperature close to thermal processing temperature. To prevent this, suitable stabilizers needs to be incorporated during manufacturing. The use of octylin compounds as stabilizer, approved by Food and Drug Administration, has increased its horizon in food packaging. Residual content of vinyl chloride monomer must be regulated as per maximum permissible levels allowance. Very clear and glossy films can be manufactured having high tensile strength and density. The higher WVTR (30 to 40 g /m^2 /24 h at 38^0C and RH 90 %) and lower GTR (300-400 cc/m^2/24 h at 25^0C) are observed as compare to polyethylene. This form has not only better resistance to oils and fats but also resistance to acids and alkalis. It is being mainly used biscuits and chocolate trays or containers. Moreover, it is also used for fruit juices and edible oil packaging.

Plasticized form of PVC properties depends upon the type and amount of plasticizer used during manufacturing. Phthalic acid esters are most commonly used plasticizer. This facilitates internal movement of molecular chains. Plasticized film is dominantly used for stretch wrapping of fresh fruit & vegetables and fresh meat. Comparatively, higher WVTR enables to escape moisture and prevents condensation inside of the film. Shrink wrapping is also possible if the film is oriented. Both, un-plasticized and plasticized form of PVC film can be heat sealed by high frequency heat sealers.

2.4.6. Polyvinylidene Chloride (PVdC)

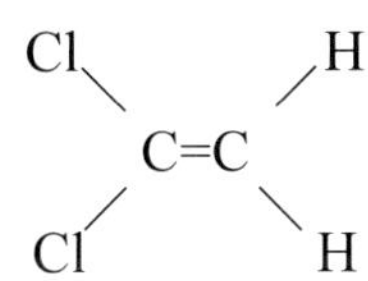

Vinylidene chloride monomer (1, 1- dichloroethylene)

Polymerization of vinylidene chloride monomer results into a very stiff film. Such film does not have application in food packaging. But when it is polymerized with vinyl chloride, a flexible, soft and highly impermeable film is produced. The company Dow named this film as 'Saran' and marketed since 1940. Shrinkable form of PVdC is also available under trademark as 'Cryovac'. Degree of polymerization and amount of VCM affects on the properties of PVdC. WVTR and GTR of 100 micron thick film are 2-4 g/ m^2/24 h at 38 ^{0}C and 5-25 cc/m^2/24 h at 25 ^{0}C respectively. This shows that it is the best barrier film among various plastic grades. Therefore, PVdC film is used in retention of food flavors and odors. This film performs excellent when coated on regenerated cellulose. It has the ability to withstand high temperature processes like retorting or hot filling. It can be used well in laminate as the most performing barrier or impermeable layer. Such laminates are used for packaging of powdered soup and sauces, powdered coffee, baby foods, dried fruits, cheese and sweets. It not only possesses excellent oil and fat resistance but also stress crack resistance. So, foods intended for distribution and microwave heating can be packed in PVdC film. Laminate combination like PE/PVdC/PE can be used for meat packing.

2.4.7. Polystyrene (PS)

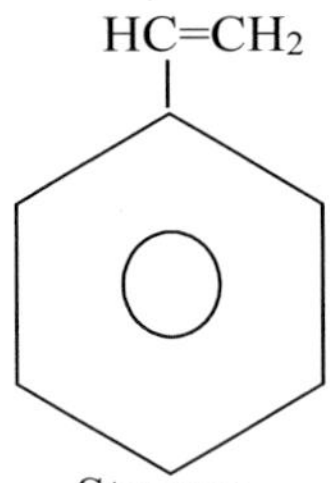

Styrene

The process of dehydrogenation is employed on ethyl benzene to produce styrene. Addition polymerization of styrene is carried out to yield polystyrene (PS). It is mainly atactic polymer with amorphous nature due to benzene ring structure. This also imparts stiffness to the material. It makes a characteristic metallic sound when dropped on a hard surface. It possesses poor barrier to water vapor (100-155 g/m^2/24 h at 38 ^{0}C for 100 micron thick film) and reasonable barrier to gas (4000-5000 cc/m^2/24 h at 25 ^{0}C for 100 micron thick film). PS co-extrusion with barrier grades like PVdC or ethylene vinyl alcohol (EVOH) produces thermoform containers for non-perishable food products. The extruded PS foam is branded by Dow Company as 'Styrofoam'. It has better insulating property. This form is mainly used for disposable packages such as parcel take away meals, dinnerware, cartons and trays. High impact PS (HIPS) grade is manufactured with inclusion of poly-1,4-butadiene. HIPS have higher impact strength and flexibility but lower tensile strength and thermal resistance.

2.4.8. Polyethylene terphthalate (PET)

PET is a polyester type containing carbon-oxygen-carbon linkages. It is manufactured by the condensation polymerization. The reaction between terephthalic acid and ethylene glycol is the basis for PET manufacturing. Practically, di-methyl esters of terephthalic acid is used to give better controlled reaction. The un-oriented film form has no application in packaging as it is extremely brittle and opaque. The bi-axially oriented and heat stabilized form of PET films are widely used. Heat stabilization i.e. annealing is carried out to make it more useable. PET film has excellent tensile strength, mechanical strength and stability. The softening point is about 250^{0}C. These characteristics led the use of this film for boil-in-bag food products and oven bags which can sustain processing temperature. It can also sustain lower temperature up to -70 ^{0}C. It has lower water vapor and gas permeability i.e. 21-30 g/m^2/24 h at 38 ^{0}C and 50-90 cc/m^2/24 h at 25 ^{0}C respectively. The outstanding printing quality makes it more favorite choice for outer layer in laminates. It has a poor sealing property. Therefore, solvent based adhesives are used in formation of this film laminates. Metallized PET films are commonly in use of laminates. Coatings of PVdC copolymer increase its barrier properties towards water vapor and gases. PVdC coated PET films is majorly used for cheese slice wrap. LDPE coated PET film is very easy to seal and still tough film can be obtained. Such films are common for packaging of snack foods. Rigid form of PET in bottle manufacturing is also very widely used. The bi-axial orientation have made possible for PET bottle manufacturing. Mostly stretch – blow molding is employed for bottle manufacturing. PET bottles are used for packaging of carbonated beverages, fruit juices or drinks.

2.4.9. Crystallized PET (CPET)

CPET is a special co-polyester grade of PET. This is manufactured by addition of nucleating material (e.g. 3-methylpentane-2-4-diol) while molding at 170 ^{0}C. The unique step makes it more crystalline and imparts final degree of strength up to 230 ^{0}C. The CPET grade is commonly being used for ovenable or micro-ovenable trays for frozen Indian curries or meals. This grade can sustain cold temperature (frozen storage below -18 ^{0}C) and also the oven temperature just before the consumption of food product. Active packaging PET grade having MxD6 Nylon plus cobalt has been reported to have oxygen scavenging property.

Overall, miscellaneous additives and manufacturing technologies are being used for inducing and controlling crystallinity, thereby improving thermal stability and impact characteristics. Bi-axially oriented PET films are being widely employed in food packaging. Inducing different states of crystallization in PET film is crucial for achieving desirable properties considering the food packaging requirement.

2.4.10. Polycarbonate (PC)

Polycarbonate (PC) is thermoplastic polyester comprising of carbonate linkages [-O-CO-O-]. At first, it was manufactured by the reaction between bisphenol A and phosgene. Presently, phosgene is substituted with diphenyl carbonate. PC has amorphous nature and density in the range of 1.2 to 1.22 g/cm^3. It offers excellent transparency, toughness, abrasion resistance and thermal resistance. It provides more freedom in designing aesthetically appealing packaging compared to other plastic grades. It is widely used owing to its high impact strength and high dimensional stability at wide temperature range. The softening point is in the range of 140 to 165°C and it can perform well at -20 °C. It has good resistance towards dilute acids but gets attacked by alkalis and bases. The barrier property is low (Water vapor transmission rate 50 to 150 g/m^2/24 h at 38 °C; Gas transmission rate 3000 cc/m^2/24 h at 25 °C). To improve barrier property, it must be coated. Bi-axially oriented PC is being used for electrical applications and not in food packaging. Generally injection and blow molded PC is being used for refillable water cans, baby feeding bottles and other reusable bottles. Importantly, PC is used for frozen foods trays which can be micro-ovenable. Moreover, PC film can be co-extruded with PET resulting into durable and toughened film. The film can be used in boil-in-bag package or retortable package due to its thermal stability at higher temperature. PC has higher resistance to staining. Owing to this property, PC is used in hotel jugs and glasses manufacturing intended for tea, coffee, fruit juices and sauces servings.

2.4.11. Polyamide (PA) or Nylon

Nylon is a common name for a thermoplastic polyamide having carbon-carbonyl-nitrogen-carbon linkage (-CONH-). This linkage is also referred as amide group. PA or nylon is manufactured by condensation polymerization. PA or nylon imitates like of silk. Du Pont company carried out pioneer work in polyamide. Nylon 6 was firstly introduced by Du Pont. Afterwards, several other nylon grades had been introduced. Nomenclature is based on a number which signifies number of carbon atoms present in di-acid from which it gets manufactured.

Table 1: Nylon grades and their properties

Nylon Grade	Manufactured from	Carbon atom presence
Nylon 6	Caprolactum	6 carbon in caprolactum
Nylon 6,6	Hexamethylene diamine and adipic acid	6 carbon in Hexamethylene diamine and 6 carbon in adipic acid
Nylon 6,10	Hexamethylene diamine and sebacic acid	6 carbon in Hexamethylene diamine and 10 carbon in adipic acid
Nylon 11	Thermally decomposed Ricinoleic acid to C11 acid (undecanoic acid)	11 carbon in undecanoic acid

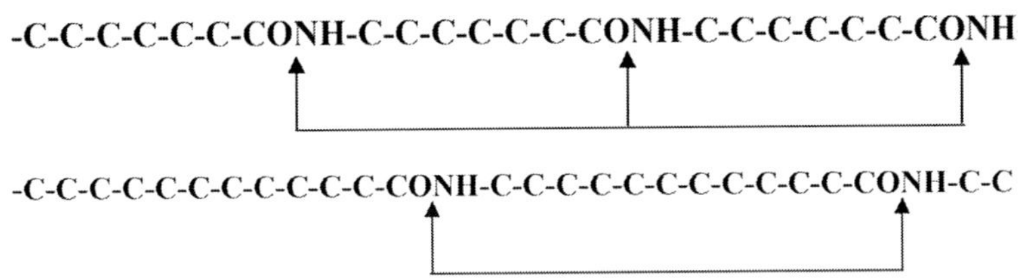

The distance between repeating –CONH- polar group can considerably affect properties of nylon. As the distance between aliphatic segment polar group (-CONH-) increases, reductions in melting point, tensile strength and water absorption are observed. However, an increase in elongation and impact strength occur simultaneously. Therefore it can be expected that Nylon 11 will have lower melting point, tensile strength and water absorption than Nylon 6 while vice versa with elongation and impact strength.

The softening point of polyamide is around 215 to 265 °C. So, it indicates that it can withstand high temperature processing. However, heat sealing requires higher temperature i.e. approx. 240 °C. It has higher mechanical and elongation strength. It has also excellent printability and resistance to cutting. Generally polyamide has low barrier property towards water vapour and gas transmission. Nylon 6 has higher water vapor transmission rate than Nylon 11 while gas transmission rate of Nylon 6 is lower than that of Nylon 11.

Table 2: Comparison of properties of Nylon 6 and Nylon 11

Polyamide (PA)	**WVTR (g/m²/24 h at 38 °C, 90% RH)**	**GTR (cc/m²/24 h at 25 °C)**
Nylon 6	250 - 350	40
Nylon 11	60 - 100	400

The permeability of PA films changes with varying relative humidity. Biaxial orientation of polyamide films provides better spectrum of mechanical and barrier properties. Moreover, oriented PA films imparts crack resistance. Such films are now commonly used in the packaging of processed cheese, fresh meat and frozen foods. Laminated polyamide films are majorly used in food packaging. PVdC coated polyamide film provides enhanced water vapor, gas and UV light barrier characteristics. Coated polyamide films are having excellent odor retention property.

2.4.12 Ethylene-vinyl Acetate (EVA)

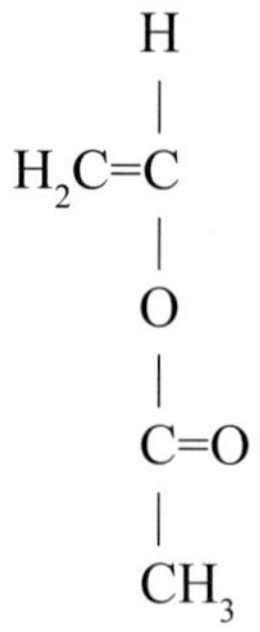

EVA is a copolymer of ethylene with vinyl acetate (VA). It is manufactured by free radical polymerization. It is similar in flexibility to plasticized PVC. Generally, EVA may contain 1 to 50% VA monomer. Due to presence of bulky polar VA, the ethylene carbon backbone can direct the end properties of the copolymer by varying and optimizing the VA content. As VA content increases, the crystalline nature decreases and becomes more elastic. Consequently, permeability to gas, water vapors, fat & oils increases. But impact strength increases. EVA copolymers are used mainly in coating, laminating and as flexible films. The co-monomer, VA, being the cheap and non-toxic, is used in food packaging applications. EVA is having very low barrier property for gas and water vapor (GTR - 10000 cc/m^2/24 h at 25 °C and WVTR – 20 to 150 g/m^2/24 h at 38 °C, 90% RH). This limitation is exploited as its strength by using in the form of stretch film or cling wrap. It can stretch up to 50%. It is being used for packaging of fresh meat particularly. Other fresh fruits and vegetables can also be packed with EVA stretch film. This film will not allow for moisture condensation inside the package and further microbial growth on the food product.

2.4.13. Ethylene-vinyl alcohol (EVOH)

EVOH is a another copolymer of ethylene produced as hydrolyzed copolymer of vinyl acetate and ethylene. It is a random copolymer constituting of ethylene and vinyl alcohol. Different grades of EVOH are commercially available depending on mole percent of ethylene. Generally, ethylene content ranges from 27 to 48 mole percent. The ethylene content of EVOH is significant for film formation and stretching without break. On other side, gas barrier property decreases and moisture barrier property improves with an increase in ethylene content. Optimized ethylene content in EVOH can be worked out to address both film or container formation and barrier properties. The outstanding property of EVOH is the capability to provide a barrier to gases (GTR–0.5 to 20 cc/m^2/24 h at 25 °C). EVOH packaging film or coating has better odor, flavor and quality retention by eliminating oxygen penetration into the package. Moreover, if the package is having gas flush (Carbon dioxide or nitrogen), it can retain effectively inside the package. Therefore, EVOH can be employed in packaging where the product is rich in aroma and flavor. EVOH is hydrophilic and can absorb moisture if stored at high relative humidity environment. If the moisture is absorbed, gas barrier properties are at the stake. However, EVOH has low haze, high gloss, resistance to oils and excellent abrasion resistance. EVOH is used in the manufacturing of bulk bags used for aseptic packaging, containers and retort pouches.

2.4.14. Ionomers

Ionomer is a polymer composed of repeated units of both electrically neutral units and ionized units covalently bonded to the polymer backbone. Generally not more than 15 % mole are ionized. Often carboxylic acid groups are the ionized units. The company Du Pont first manufactured ionomer by copolymerizing ethylene with a small amount of methacrylic acid (carboxylic acid) under high pressure. Thereafter, this copolymer was neutralized with sodium methoxide and magnesium acetate, causing the carboxylic acid to ionize. This confers enhanced stiffness and toughness to ionomer film. This film is tougher and transparent than PE film, also has excellent fat & oil resistance and high puncture resistance. It can be used for aseptic packaging of fruit juices and snack product containing fat. Ionomer coextruded films are used in laminations and extrusion coatings in major flexible packaging materials. These include paper composite based containers/pouches, form fill seal packaging materials, controlled and modified atmosphere packaging materials. It has also excellent sealing strength, so high speed packaging machines enables to run smoothly. Ionomers can be coated or laminated with other plastic grades like PE, PP, PET, PA and also with aluminum foil.

2.14.15. Retortable/ Micro-ovenable Flexible Packages

Modern urban society is trending on the fact ‘Cash rich, time poor’. The beginning of this life style paved the way for ready to eat (RTE) and ready to cook (RTC) food products market. Already several food products (RTE or RTC) are showing substantial market potential and occupied the spaces in retail or hyper market. Keeping this mind, convenient and easily disposable packaging materials are the need of this food category.

Retortable packages are available as composite flexible pouches and retortable containers. These packages are sterilizable. They can be used for viscous

Indian curries packaging. The consumer can heat the product in pouch (Boil-in-pouch) itself or empty out the product in utensil and then heat. The product is ready for consumption after heating. Micro-ovenable packages are generally manufactured from special aluminum laminate which can withstand microwave heating and processing treatments.

Functional layers in flexible laminated packages

Functional layers can be classified on the basis of their functions as follows.

- Structural layer
- Barrier layer
- Adhesive layer
- Sealing layer

***Structural layer*:** This layer is significantly important in flexible packaging material for imparting body, strength and overall structure to the package. Plastic grades like PP, PET, Polyamide can be used for this purpose. Other than plastic based laminates, paper & paperboard are the common structural layers in tetra packs.

***Barrier layer*:** This layer is vital in providing barrier properties in terms of water vapor and gas vapor permeability resistance. Plastic grades like BOPP, EVOH, PVdC can be used for this purpose. Other than plastic, aluminum foil is the most commonly used as barrier layer in most of the common laminates.

***Adhesive layer*:** This layer is responsible for holding two different layers together. It should have ability to make good adhesion and instant tack. LDPE is the most common adhesive layer used in the plastic or paper or aluminum based laminates. Moreover, PP is best suited for specific retortable packages. Ethylene acrylic acid (EAA) and Ethylene metacrylic acid (EMAA) can also be used as adhesive layer in laminates. The carboxyl groups of EAA or EMAA provide a site for hydrogen bonding between molecules. The final laminate is tough, tear resistant good hot tack and low temperature sealability.

***Sealing layer*:** LDPE is the most favorite choice for sealing layer as LDPE is having low softening temperature (85-87^{0}C), still provides instant tack after cooling. Moreover, sealing strength is better. Ionomer can also be used as sealing layer.

Retortable flexible pouches can be manufactured in flat configuration and stand-up configuration. Typical laminate configurations are as follows.

Type	External Side	Middle	Product Side
Flat configuration -	12μ PET/	12 μ Al foil/	75 μ PP
Stand-up configuration -	12μ PET/	9 μ Al foil / 15 μ OPA/	60 μ PP

Simple flexible pouches can also be manufactured in flat and stand up configurations with the appropriate plastic grades and other packaging materials like aluminum and paper.

The process of lamination is holding two or multiple packaging layers together by either use of adhesive, heat or extrusion process. Three major different types of lamination are used in food packaging. The selection of the most suitable lamination process is based on the end-use of the product.

Wet lamination

Wet lamination process is used to laminate two layers, one of which is porous or permeable to solvent used for bonding. This process uses aqueous adhesives such as casein, sodium silicate, starch, PVA latex, rubber latex or dextrin. As the wet bonding process is applied to each substrate/layer, the two substrates get combined prior to passing through an oven, which is used for the drying process. During drying process, the adhesive materials dries up, essentially adhering the two layers together. Therefore, the two substrates need to be combined before the adhesive is dried in the oven. Paper and aluminum foil lamination is the most common example of wet lamination. It is not successful with plastic films.

Dry lamination

Dry lamination process is used to combine two or multiple non-porous or impervious layers. The adhesive used to combine the two layers together is applied to one of layer, and then dried in an oven. The adhesive coated first layer is then laminated with the second layer under strong pressure and using heated rollers. The two layers are combined after the adhesive is dried for maximum efficiency of the dry bond lamination. Plastic film on other plastic film or aluminum foil can be laminated with dry bond lamination.

Extrusion lamination

Extrusion lamination and coating has excellent diversified potential in combining multiple flexible packaging films. It offers an opportunity with great flexibility to develop innovative and proprietary packaging materials. Solid granules of selected polymer are feeded in the extruder and molten polymer flows through the die slot. A single polymer or a blend of polymers can be selected. The combined set of multiple extruders can be arranged to develop different layered configuration (generally referred as co-extrusion). Stronger adhesion results either when atoms of two materials exchange electrons (ionic bonding) or share the electrons (covalent bonding) (Dunn, 2015).

Advantages of retortable flexible laminated packaging materials

- Convenience in use
- Almost 100% barrier and so better product preservation
- Excellent shelf life
- Flexibility in size and shape of packages
- Optimum processing line suitability
- Effective package space utilization for advertisement as flat and amenable printing surface area can be developed.
- Gentle temperature can achieve pasteurization or commercial sterilization due to higher heat transfer compared with glass and tinplate containers.
- Brine or other heating medium are not necessary
- Low weight and storage space requirement as they have good stackability.
- Less transportation costs
- Easily recycle with low energy cost
- Economical

Limitations of retortable flexible laminated packaging materials

- Flexible pouches are more prone for puncture.
- High strength secondary packaging materials are necessary.
- Different layers selection and combination for a well performing laminate packaging material requires sound knowledge about the different plastic and other packaging materials.
- Has adverse environmental effect
- It can absorb flavour and smell of food material.

References

Dunn Thomas (2015). Extrusion lamination and coating. In Manufacturing flexible packaging: Materials, Machinery and Techniques. PDL Handbook series by Elsevier. pp: 49-59.

Fellows P. (2000). Packaging. In Food Processing Technology: Principles and Practice. pp: 462–510. Woodhead Publishing Ltd., Cambridge, England.

FICCI (2014). Potential of Plastics industry in northern India with special focus on plasticulutre and Food Processing – 2014. A report on Plastic Industry. By Federation of Indian Chambers of Commerce & Industries and Department of Chemical & Petrochemicals, Government of India. Downloaded from https://ficci.in/spdocument/20396/knowledge-paper-ps.pdf

Mukherjee S.B. (2006). Packaging: An Overview. Packaging India. Oct.-Nov. 2006. pp: 31-48.

National Institute of Industrial Research (NIIR). Food Packaging Technology Handbook by NIIR, Delhi Board of Consultants and Engineers. Published by NIIR, Delhi, India.

Paine F.A. & Paine H.Y. (1992). A Handbook of Food Packaging. Second Edition. Blackie Academic and Professional, London.

Robertson G.L. (1993). Food Packaging: Principles and Practice. Marcel Dekker Inc., New York.

3

Paper and Paper-based Packaging

Srishti Upadhyay, Gourav Chakraborty, Tanmay Yadav and Vijay Singh Sharanagat

Department of Food Engineering, National Institute of Food Technology Entrepreneurship and Management, Kundli, Sonipat, Haryana, India

3.1. Introduction

Paper and paper-based packaging is the traditional form of packaging used for various food products due to its various advantageous characteristics. It is less expensive, flexible, lightweight, durable, recyclable and available in various shapes, sizes and colour to meet customer needs. Hence it occupies a vast market in food packaging from primary (i.e., packaging which is first used after manufacture) to secondary (i.e. processing aids of primary packages) packaging and the production of paper has doubled in the last three decades. In the year 2020, china was the leading producer for paper and paper based packaging followed by USA and Japan (Figure 1).

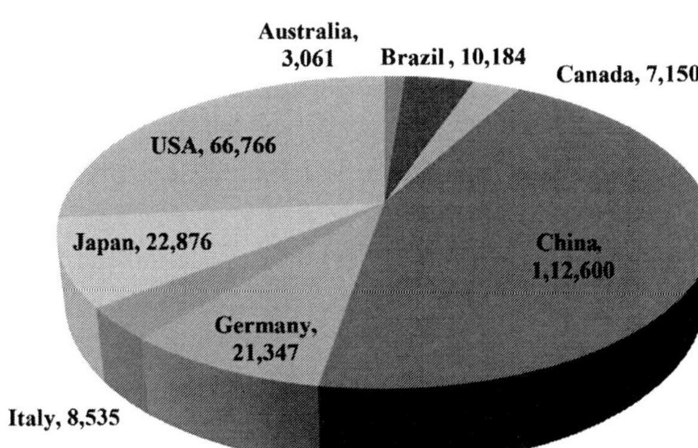

Fig. 1: Paper and Paper board production by different countries in the year 2020 in hundred tonnes (*Source*: FAO, 2020).

Paper is a sheet of cellulose network arranged to form a material considered for primary packaging. It is traditionally made from the wood using mechanical or chemical method. Paperboard is a secondary form of packaging used for packaging food items in bulk and has a higher capacity/strength compared to paper. The paperboard is thicker than paper with more weight grams per square meter. Wood pulp plays an important role in paper and paper-board manufacturing as 97% of the world's paper and board is obtained from cellulosic fibers of wood pulp while about 85% of the wood pulp used is obtained from trees of spruces, firs and pines-coniferous that predominates in the forests of the north-temperate zone.The manufacturing process of paper and paperboard is similar and mostly single Fourdrinier wire machine is used. According to the requirement, various types of boards are manufactured which can be single sheet boards or Multi-ply boards. They are generally used for manufacturing boxes, folding cartons, beverage cartons, etc. Multi-ply boards are basically used in manufacturing rigid boxes, folding cartons, and beverage cartons and are made of multiple plies combined to form a single sheet of paper. Waste material like old newspapers can be used to manufacture multi-ply boards. Waste material can be incorporated in inner layer of boards where not much strength is required. However the waste material having extraneous material may causes a toxic effect on the packaged material (Smook, 2015). Generally, these multi-ply boards that contain toxic waste cannot be used for food contact purposes. With the rising concerns of global warming and depletion of natural resources, paper is now also made by used paper called recycling of wastepaper.

Paper packaging can be parchment paper or the paper that is used for packaging of loose foods, carton boxes used for large-scale packing as well as packaging liquid and dry foods, frozen foods and fast food, corrugated boards used for packaging food such as pizza (in primary packaging) and in some places secondary packaging also. Beyond the conventional packaging paper and paper boards are now also used for ice-cream cups, juice containers, chocolate packages, popcorn containers, bakery cups, juice cartons, and ready-to-eat containers such as pizza boxes and tea cups. Paper and paperboard are unique form of packaging which can be used for all types of food products. Compared to many advantages it has some limitations too as it is highly permeable to the aqueous solution including water, water vapour, organic solvents, in some cases gases, aggressive chemicals, volatile flavours and aroma. These limitations can be overcome by adding specific additives to the packaging material, which increases the heat sealable and barrier properties, increasing the effectiveness of paper-based packaging.

3.2. Manufacturing of Paper

Paper and board are manufactured by natural fibers (virgin fiber) of wood that can be bleached or it can be obtained from waste material (used paper and paperboard). Many chemical additives are also added to impart desirable strength and functionalities. Additives are sometimes added at the stock preparation stage and sometimes coated on the surface at latter stages. Production of paper is a two-step process: the first step is the production of pulp from wood and then the conversion of pulp into paper (Ververis et al., 2004). Paperboard is also manufactured by same process with higher grammage value. Manufacturing process includes harvesting the forest and debarking of logs, chipping and pulping, refining, sheet forming, drying, calendaring and winding (Figure 2). Details process of paper manufacturing has been described in subsequent sections.

Fig. 2: Processing of paper

3.2.1. Manufacturing of Pulp

Manufacturing of pulp is the first step towards paper manufacturing as it involves production and processing of raw material. Wood is received in different forms based on the requirement of pulping process and on the type of wood. It can be bolts (short logs) of round shape with barks of trees on the surface or it can be small wood chips that are grinded from the saw mill. Bolts are required to be debarked and chipped for pulping process. Pulping is a multi-stage process which aims to remove lignin and impurities from wood without losing fiber strength. It involves crushing, washing, screening, bleaching, cleaning and drying plant fibers. The steps can be altered depending upon the type of pulp to be produced. Pulp with longer fibers and less lignin is considered to be good choice for producing the strongest paper. Mechanical and chemical pulping has been extensively used methods to separate fibers from lignin (Figure 3). In mechanical pulping, logs or wood chips are converted into pulp by using mechanical grinding whereas in chemical pulping chemicals are used to dissolve and remove the lignin from the cellulose fibers (Figure 4).

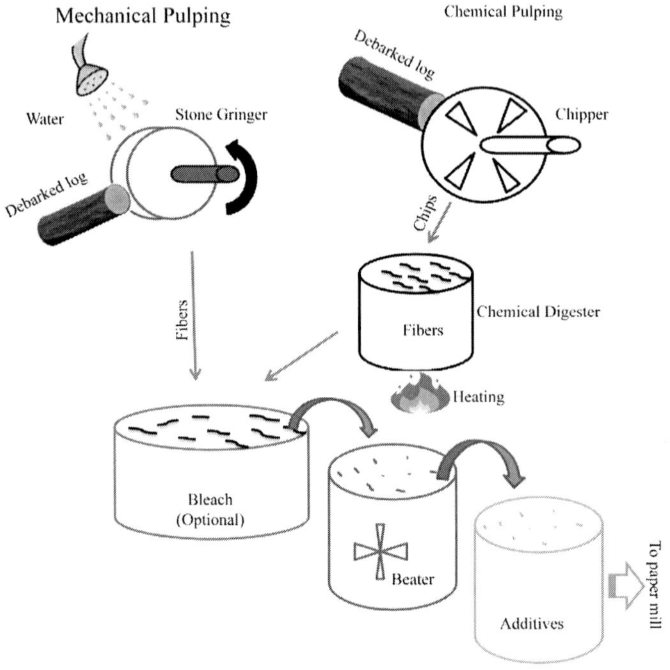

Fig. 3: Mechanical and chemical pulping steps.

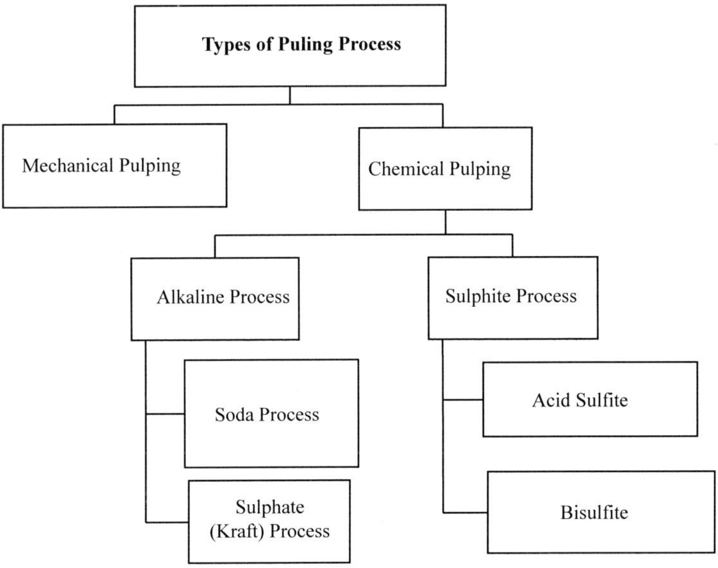

Fig. 4: Types of Pulping Process

3.2.1.1. Types of Pulping Method

(i) **Mechanical Pulping:** Mechanical pulping is a traditional method which grinds wood logs by abrasive action of revolving stone. In this method of pulping wood logs are forced into the revolving stone and water is sprayed to prevent fiber damage due to frictional heat and for the removal of fibers from the wood. Now these days, Refiner Mechanical Pulping (RMP) method is used which provides much greater strength to pulp compared to traditional method. It sandwiches wood chips between revolving disks which generate heat due to friction producing more soft lignin with greater separation of the cellulose fibersand comparatively less fiber damage. Hence the RMP involves little or no need to be supplemented with chemical pulping as it already involves more strength to fibers.Thermomechanical pulping is another form of pulping similar to RMP producing more fiber separation, requires higher temperature and pressure to soften the lignin. De-fiberization can be increased in thermomechanical pulping by increasing the passes producing much stronger pulp and more yield at same cost compared to RMP. Chemi-thermomechanical (Semi-chemical) pulping is another method used similar method of pulping but involves use of mild chemicals which increase pulp brightness and reduce shive content. It also involves the use of pressurized ground wood which operates under high pressure which increases the pulp yield and more fiber separation.

(ii) **Chemical pulping method:** Chemical pulping i.e kraft pulping and sulfite pulping can be characterized by the use of chemical for pulping. Kraft (Sulfate) pulping is most common pulping method used by 80% of the chemical pulping industry. It involves digestion of wood chips at higher temperature and pressure in the solution of water with sodium sulfide (Na_2S, also named as white liquor) and sodium hydroxide as pulping agents. The digestion is achieved in either a batch digester or a continuous digester. Whereas in Sulfite pulping as the name suggests, it involves sulfite or bi-sulfite salts of sodium, calcium, potassium, magnesium, and ammonium to remove lignin from wood by converting it into lignosulfonates, which are easily separable from cellulose fibers. It produces stronger fibers, with less harmful effect to environment compared to kraft process. Table 1 summarized the different characteristic of pulp produced by mechanical and chemical pulping method.

Table 1: Difference between Mechanical and Chemical Pulp

Properties	**Mechanical Pulping**	**Chemical pulping**
Yield fiber/wood	High 90-98%	Low 40-70%
Cellulose purity	Low-lignin remains	High-lignin dissolved
Final produced paper	Low quality High volume paper	Dissolving pulp High-quality paper
Raw material sensitivity	High	Low
Strength	Low - fibers damaged	High - fibers intact
Bulk	High - few and less flexible fibers	Low - more and flexible
Optical	Bright but hard to bleach high Good light scattering	Dark but bleachable Poor light scattering
Drainability	Poor - short fibers	Good - long fibers
Permanence	Poor (optical)	Good
Example	Newsprint Molded products	Book, Reinforcement package

3.2.1.2. Digestion

Digestion is the process to free lignin and other non-cellulosic content and require one and a half hours at 140-180°C at about 10 atm. The pulp cooking process can be divided into batch and continuous, according to the difference in working mechanisms. In addition, vertical digesters and horizontal digesters are two types of most common digesters used in chemical pulping productions. The vertical digester is used mostly for super batch cooking. Although, increasing performance of a digester may directly affect paper quality and quantity but also have a negative influence on environment, more operating cost and production losses.

3.2.1.3. Beating and Refining

The beating and refining are the mechanical treatments of chemical pulps which aim to increase the inter-fiber bonding by swelling fibers in water and increasing surface area. These treatments develop the bonding ability of fibers without affecting the individual strength of fibers. Beating and refining is a random and repetitive application of tensile, compressive, shear and bending forces which leads to respond in these ways:

(i) It leads to development of new surfaces externally through fibrillation and fiber wall delamination internally.

(ii) It also leads to deformation of fibers resulting in change in geometric shape, fibrillar alignment along the length. It also leads to flattening or collapsing of fibers.

(iii) It is also observed that fibers break after beating and refining leading to changes in length distribution with decrease in mean fiber length.

Various additives such as alum or synthetic resins are also applied in beating to increase the water-repellency of the fibers. They also improve the wet strength of the paper. Other agents like Fluorescent whitening agents (FWAs), commonly known as optical brightening agents (OBAs), are also used for increasing whiteness and brightness.

3.2.1.4. Bleaching

Raw pulp contains number of impurities, with considerable amount of lignin and other discoloration which needs to removed. Bleaching is a chemical treatment given to chemical and mechanical pulps, for increasing the brightness, removing discoloration, giving an appreciable look to pulp by removing extractives and other contaminants. Cellulose and hemicelluloses are naturally white and hence do not contribute to color, while chromophoric groups which are present in lignin contribute for the color in pulp. Unbleached pulps posses less brightness value whereas bleaching reduces the strength of paper. Hence, an appropriate treatment should bc provided which gives a finite value for necessary brightness and tensile properties of paper.Some common bleaching agents which are commonly used are chlorine dioxide, chlorine gas, sodium hypochlorite, hydrogen peroxide, and oxygen to delignify additional lignin by solubilizing through chlorination and oxidation.

Table 2: Summary of different Pulping Process

Pulping Process	Mechanism for pulping	Pulp Yield	Pulp Properties	Application
Mechanical Pulping	Mechanical Forces	High(85-95%) lignin Preserved	• Short, weak, unstablefibers • High Opaque Paper • Good Print Quality	For production of Paper such as Newspaper, Magazine, Books, Container board
Chemical Pulping	Chemical Reactions	• 70% yield from brown paper • 45-55% yield from bleached paper	Long, strong stable fibers with rich quality and color	Kraft bags, wrapping linear board, newsprint, graphics, writing paper. Sulphite paper: Fine paper, tissue, glassing, newsprint
Semi-Chemical	Combination of Mechanical and Chemical Pulp	Yield is 55-85%	Pulp properties are similar to both mechanical and chemical pulp	Corrugated Board, food Packaging
Recycled	Mainly mechanical and some heat and chemical	Depends on waste paper source	Mixture of fiber grades properties depend on fiber grade source	Newsprint, Magazine, packaging, Tissue and writing paper

3.2.2. Pulp to Paper Manufacturing

Pulp preparation is first step of paper making and final paper can be prepared by further processing of pulp using the different steps (Figure 5). Stock Preparation/stocking is the first step in paper making process as pulp is transported from pulp mills to paper making mills. Sometimes the produced pulp is treated chemically after pulping to make it ready for sheet forming. In stock preparation, the pulps are made by dissolving slurries in water however in some cases drysheet of pulps are dispersed in water for re-pulping. Hydra-pulper is commonly used for this purpose. When the pulp mill and papermaking are close to each other, re-pulping is not needed as direct pulp is delivered to the paper making mills. Pulp obtained from pulp making mills is used to form sheets. In this process pulp is laid on the plates to form long sheets of paper. The pulp sheets are transported to pressing and drying section using the long moving belts. Pressing and drying is connected process wherein first the papers are pressed in the pressing section and then it is dried using dryers. In the pressing section, it is passed through a series of rolls which squeezes all of the water by compression. The amount of water removed in pressing sections depends much on the pressure applied by machine as well as running speed which reduces the moisture content up to 65%. Rotary presses are another category of presses having solid or perforated rollers use internal suction for pressing the between the rollers. After pressing the papers sheets are passed the series of steam-heated rollers which dries the sheet up to the moisture content of 4 to 10%. Pressing and drying is considered as the final step in paper manufacturing. Once the paper sheets have been developed, different operation are performed for finishing and preparation of different types of paper.

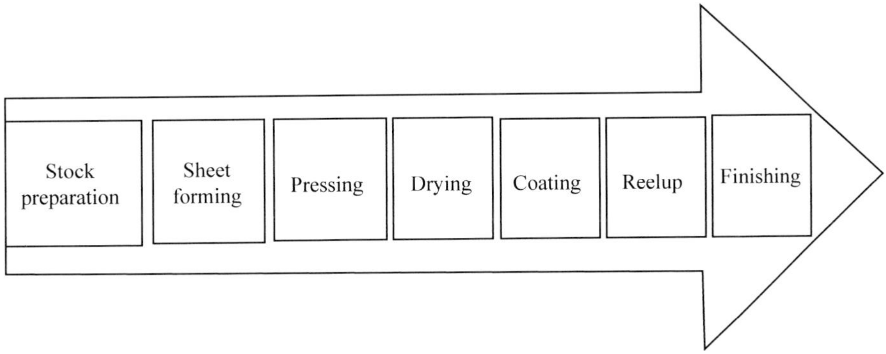

Fig. 5: Steps in pulp to paper and paperboard manufacturing.

Different processes used for converting and finishing are given below-

(a) **Calendering:** Calendering is an on-machine smoothening process to produce even and smooth paper by passing the paper through a series of steel rolls before rewinding on a reel. The rolls in combination are called calenders. Calenders produce a high-density sheet with less variation in sheet with optimum smoothness. It can also affect the water absorption properties of the paper.

(b) **Sizing:** Sizing can be achieved from surface-sizing agents to prevent the excess water penetration and improves the strength of paper. Sizing agents penetrates into the paper and increases the fiber bonding and directly impact the bursting, tensile and folding strengths of paper. Surface sizing in paper can be achieved by starches (generally chemical modified) such as oxidized starches, cationic starches and hydroxyethylated derivatives. Film forming materials such as carboxymethyl cellulose (CMC) and poly (vinyl alcohol) (PVOH) are also used for surface sizing that improves the paper strength and decreased porosity and also work as oil and grease-repellent coating.

(c) **Barrier Coatings:** Coating is secondary unit operation in paper manufacturing process which enhances paper printing quality, color, smoothness, opacity and other surface characteristics by use of clay or other pigments and/or adhesives. "Coated free sheet" are the paper sheet made from ground wood-fibers produced from chemical pulp. Different classes of coated paper like glossy, dull, and mat are available.

Air knife and blade coating are popularly used coating methods. In Air knife coating process, a jet of air is used to remove excess coating from the paperboard to provide surface finish and even coating on the surface. Blade coating process is simple process which uses flexible blade that is set at an adjustable angle to remove excess coating across the web. Following the coating operation, the sheet must again be dried and rewound.

Paraffin wax coating is an age-old coating method is commonly used as water repellent coating. Nowadays wax is being largely replaced by LDPE applied by extrusion coating producing a more durable, flexible coating that can be heat sealable to itself and are removable in hydrapulper. Paraffin wax has major limitation in recycling the paperboard as wax cannot be separated from the fibers so that they can be re-pulped. Some bio-based polymers which can replace paraffin wax are starch and cellulose derivatives, chitosan, alginate, wheat gluten, whey proteins, polycaprolactone, PLA and polyhydroxyalkanoates.

(d) **Pigments:** Pigments are substance used to mask or change the appearance of base stock, impart a smooth, receptive surface and improve opacity and provide special properties (color) to the paper. These usually comprise 70%-90% of the dry solids in paper coating. Kaolin clay is the highly used pigment used in paper industry, followed by calcium carbonate. These are added at the size press. Additionally, fillers are another substance similar to pigments however they are used to fill the sheets and are added at the wet end of paper.

(e) **Lamination:** In the process of lamination, multiple layers are held together with adhesives using particle or fiber-board as a base material with kraft paper as absorbent on surface. Inert resin such as melamine, are used to form a hard composite layer on the surface. It is achieved by usage of coating machines, machine presses and calendaring equipment. Lamination improves the strength, provides insulation and gives appealing appearance to the paper. It is also used for making paper thicker and stronger.

(g) **Printing:** Printing is a mechanical process to imprint the text and images by applying ink to paper using printing technologies. Printing can be divided into two types based on contact with paper.

a. Direct contact printing - In direct contact printing method, ink plate is directly applied on the paper surface.

b. Indirect contact Printing- It is a two-step process, the ink is first applied to the blanket cylinder from the plate and then printing medium comes in contact with the blanket.

3.2.3. Papermaking Machine

Three types of paper machines are generally used in paper manufacturing process:

(a) Fourdrinier machines

Fourdrinier machine is a first continuous machine used to produce rolls or webs of the paper, paperboard and fiber-board (Figure 6). It was invented in year 1799 in France by Louis Robert and that was further improved and patented by Henry and Sealy Fourdrinier. The machine consists of an endless moving belt made of perforated screens made of plastic or wire that removes excess water from the pulp forming a continuous sheet of paper for further drying by suction, pressure, and heat.

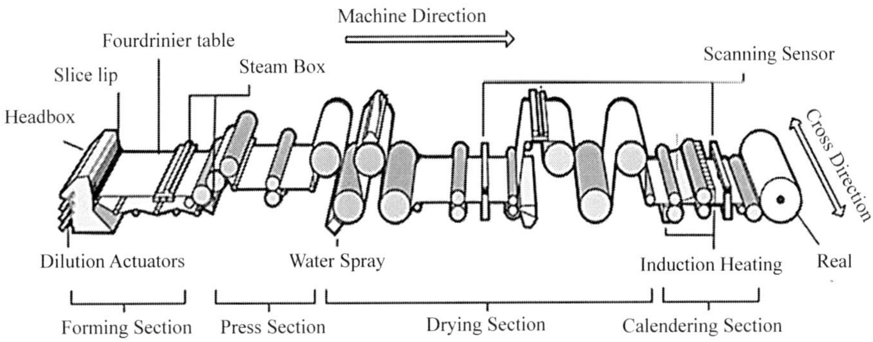

Fig. 6: Section and Components of Fourdrinier-type paper machine. (*Source*: Chu et al., 2011)

(b) Twin-wire formers

This is a modified version of a Fourdrinier-type paper machine having two wire mesh belts instead of one to form the pulp into paper.

(c) Multi-ply formers

Cylinder formers are the common type of multi-ply former, extensively used in paperboard production. It consists of a series of rotating screen-covered cylinders inside vat of dilute paper stock. As the pulp gets deposited on screens, web formation occurs on the screen as a result of suction inside the cylinder, which removes the filtrate.

Paper has a definite grain caused by the greater orientation of fibers in the direction of travel of the paper machine and the greater strength orientation that results partly from, the greater fiber alignment and partly from the greater tension exerted on the paper in this direction during drying. The grain direction is known as the *machine direction* (MD), while the *cross direction* (CD) is the direction of the paper at right angles to the MD. Papers vary in MD:CD strength ratios, with cylinder machine papers having a higher ratio than Fourdrinier papers, the latter values varying from about 1.5 to 2.5.

3.3. Types of Papers

Different types of papers are available based on production or availability of fiber, virgin or recycled fiber and whether it is bleached or unbleached etc. The amount of fiber and its processing method plays an important role in its manufacturing process. It is expressed in different forms like weight per unit area or grammage (grams per square metre, g/m^2, or lbs. per 1000sq. ft), thickness (microns, µm or 0.001mm, and thou˙(0.001 inch), and appearance (colour and surface finish).

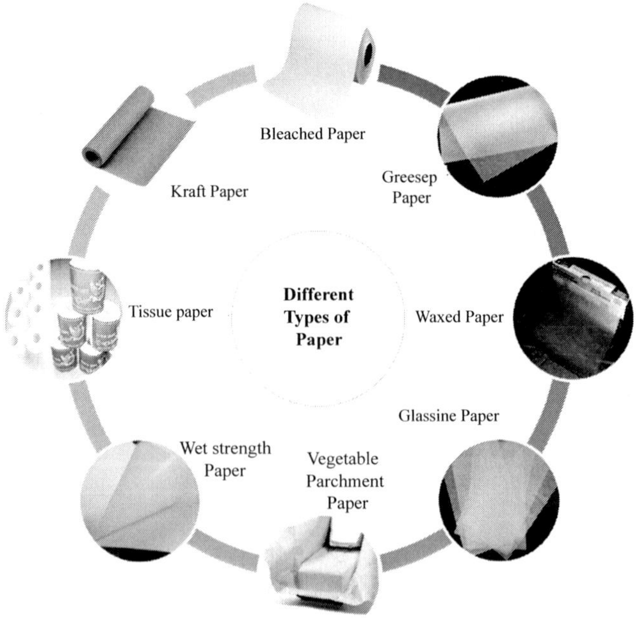

Fig. 7: Types of Paper

Papers are also commonly divided based on their properties like grammage and thickness. The thick paper can be used for heavy materials which require more loading capacity, whereas the material with less grammage value and high permeability can be used for membrane purposes such as tea bags. Surface finish is also one of the criteria for the variation in paper and paper boards. Food materials with more oil content require more surface finish to resist oil whereas the dry foods like dry fruits, flour or fruits don't require much surface finish; they just need protection. The surface finish of packaging material can be varied mechanically and the properties of the packaging material can be changed depending on the requirement by using additives at the stock preparation stage only. Some other characteristic like heat sealing properties, printing ability, strength, water resistant properties, cost-effectiveness, etc. also play an important role for the selection of packaging material. Different types of papers (Figure 7) used in different food industries are discussed below.

- ***Kraft Paper*:** Kraft paper is a brown color paper having high tensile, burst, tear and wet strength and high tensile energy absorption (Netramai et al., 2016). It is the most commonly used paper for packaging purposes. Its fibers are bleached with different bleaching materials to make different papers. White colorkraft papers are also available in the market, which are bleached to remove their color. Kraft Paper is considered as the base material for all other types of paper. It is also used to produce the liner of corrugated board which is made by the sulphate pulping process. The weight of paper ranges between 8.2 to 90.7 kg per 1000 square ft and commonly used paper weight varies from 11.3–36.3 kg per 1000 square ft (Twede, 2005). Fourdrinier machine with Yankee dryer for drying is used for Kraft papers preparation. Its major application is found in making inner plies of multiwall sacks, base paper, industrial bags, grocery bags, or plain wrapping paper which needs higher strength, printability, and appearance. Some other forms of kraft papers are also available in the market having coating or lamination to improve barrier properties and to increase strength. Kraft paper also has some additional surface properties as its surface is little bit coarse which helps stacks formation of packaging bags. For coarser touch calendaring is not done to make its surface rough.
- ***Bleached Paper*:** Bleached paper is a white color paper, bright and soft in texture, made from pulps with added bleaching agents to develop required functional properties, making it expensive. It has aesthetic appeal and is weaker than unbleached paper in terms of strength. These papers have a grammage value generally above the 134 g/m^2, typically varied in the range of 200 g/m^2 to 500 g/m^2. It is generally used for cups,

plates, printing boards, tag stock, computer cards, file folders, and index cards.

- ***Greasepro of Paper*:** Greasepro of paper is a type of paper that appears translucent and slippery with small pores and is used mainly for materials with high oil content. It is basically used in the packaging of butter. It is a low-porous paper made by refining the paper stock and then creating a sheet. The sheets are passed through hard pressure rollers or supercalendered to create a dense paper which is called Glassine. Substances like starches, alginates and carboxymethyl cellulose (CMC) are used to fill pores in glassine paper and to make it fat-repellent. These papers have a grammage value of 30–50 g/m^2. Good quality grease proof papers have all the pores filled and are used to resist fat for a long time.
- ***Glassine Paper*:** Glassine paper is the finer form of grease proof paper as all the pores in glassine paper are closed and also called super calendered grease proof paper. It is basically a brown color paper, glossy in appearance, coated with PE or a touch of aluminium foil that gives it a metallised look. Titanium oxide can be used to make it opaque or to reduce its transparency. Transparent glassine paper can be made by altering the degree of hydration of pulp and also changing the basis weight of paper. It is used for packaging fat, including materials like butter and also plasticised with glycerine over paperboard to increase their barrier properties. Its grammage value ranges from 30 to 80 g m^{-2}.

  It has high density and thickness, indicating high tensile strength, folding endurance, and stiffness compared to other thin papers. Other than advantages, glassine paper also have some disadvantages. Glassine paper is slippery and transparent which make it unsuitable for printing. In addition to that being less porous, these papers don't absorb the ink and drying also takes much more time than normal paper. It increases the risk of ink smearing, and design being damaged is higher with low printability.
- ***Animal Parchment Paper*:** Animal parchment paper is a type of writing paper/material made from specially prepared untanned skins of animals like sheep, calves, and goats. It has been used as a writing medium for over two millennia. Vellum is a finer quality parchment made from the skins of young animals such as lambs and young calves.Parchment has traditionally been used instead of paper for important documents such as religious texts, public laws, indentures, and land records as it has always been considered a strong and stable material.

- ***Vegetable Parchment*:** Vegetable Parchment paper is another form of paper that looks identical to animal parchment paper. However the specific characteristics like odourless, tasteless, and lint-free make it different from animal parchment paper. Vegetable Parchment paper is cellulose-based paper manufactured by a complex process in which rich quality web made of un-sized chemical pulp is passed through concentrated sulphuric acid which swells the fiber and fill the empty voids. It shows high water, grease and air resistance. It shows excellent strength in wet condition too and the strength can also be increased by thorough washing with water followed by drying.
- ***Waxed Paper*:** As the name indicates the paper coated with wax is known as wax paper. Waxed paper can be made from any type of papers and mainly used for protection against the vapour and moisture and heat resistance. It can be wet-waxed, dry-waxed and wax-laminated based on the manufacturing process. It is manufactured by using a shock chiller on the wax-applied paper which gives a surface finish on both sides of paper. This particular paper is called wet-waxed paper. Another type which is dry-waxed is produced by passing the wax-applied paper through the hot rollers, but these don't have a continuous film on the surface. Special treatment can be given to this type of paper to improve its adhesion properties by adding resins or plastic polymers. Thermoplastic is the latest technology that is replacing wax for this purpose.
- ***Wet strength Paper*:** Wet strength is a type of paper designed to be used in wet conditions. These papers are uniquely design to retain at least 30% of their strength in saturation (water) condition. Formalaldehyde and melamine normally are used to impart wet strength as these are deposited on the surface of cellulose fibers, making them more resistant to water absorption. Generally available wet strength paper have grammage value of 60-120 gsm.
- ***Microcreeping Paper*:** Microcreeping paper is designed to resist multiple and dynamic stresses during different operations. Apart from this, these papers are highly stretchable, with 7% stretching in the machine direction though only 2 % is normally allowed. Microcreping is made by the Clupak process which produces an invisible crimp in the paper increase the elasticity.
- ***Tissues Paper*:** Tissues papers are the most common type of paper and have many applications in the food industry. It is one of the lightest paper (Grammage, 17 to 30 g/m^2) and highly permeable to water and oil. It is manufactured by adding low chloride and sulphate residue to

neutral pH tissue. More additives can also be added to the boiling pulp to increase its strength. It is used for making tea bags, coffee bags and also used for packaging of heat-sealable products (as it contains a proportion of polypropylene fibers) as well as for non-heat-sealable products.

- ***Paper Labels*:** Paper labels have glossy and finished surfaces. Paper labels can be produced using a variety of methods, including machine glazed (MG), machine finished (MF), or calendered kraft papers (100% sulphate chemical pulp) which produces paper with the grammage of 70–90 g/m^2.
- ***Bag Papers*:** Bag papers are similar to kraft papers and as the name suggests, it is used for making bags. These papers are made using various procedures as they can be blended with recycled paper or made of 100 % recycled paper. These papers can also be made from coated or uncoated bleached kraft having a grammage of 90–100 g/m^2. These papers can have an MG or a ribbed finish.
- ***Sack Kraft*:** Sack kraft is unbleached natural paper with both sides' rough matte having grammage of 70–100 g/m^2. It has high strength and is used for medium and large size packages. Its base pulp is unbleached pulp, however, some sack kraft are made of bleached pulp too. This type of paper can also be used as a base for all types of coatings, laminating and packaging.
- ***Impregnated Papers*:** As the name suggests, the impregnated paper has high water-absorbing properties prepared from natural cotton linter and wooden pulps. These papers normally have thickness in the range of 0.5 to 1.0 mm and grammage in the range of 300–500 g/m^2. Impregnated paper with various modifications like wax coated paper is used as grease/fat resistance paper.
- ***Laminating Papers*:** Laminated packaging paper is made of aluminium to increase its appearance and water resistance. It is normally made of pulps of sulphite and sulphate and coated with a layer of aluminium foil on it. It is used for packaging snacks, ready-to-eat food, beverages, seasoning and many more. This type of paper has a grammage value of about 40–80 g/m^2.
- ***Solid Bleached Board* (SBB):** Solid bleached paper is popularly known for its usage in creative design on a good finished surface. It is produced by bleaching the pulp and then coating mineral pigments on the top surface for a surface finish. These papers have good scope for creative

designs and can be cut, creased, folded, glued and embossed freely. These papers are virgin pure papers, making them more cost-effective and more reliable for usage in food products. These are also used for aroma and flavour-sensitive products. SBBs are commonly used in the packaging of re-heatable and liquid packaging. Grammage of SBB varied from 200 to 450 GSM.

- ***Solid Unbleached Board*** **(SUB):** SUB are brown color papers made from unbleached chemical pulp and are also called solid unbleached sulphate. These are highly durable and are used where high strength is required. These are also used for packaging liquid materials to give a strong base. It has high tear and puncture resistance and represent high wet strength. These papers can be coated with mineral pigments to make them more customer-appealing. Grammage of SUB varies from 150-400 GSM.
- ***Folding Boxboard*** **(FBB):** A folding boxboard is basically a multilayer paper. It is yellowish in color and sometimes white also when the white mineral coating is applied. It prepared by combining the mechanical and chemical pulp, wherein chemical pulps are sandwiched between the mechanical pulp. FBB is made more from virgin fiber with consistent purity, so it can be considered safe for aroma and flavour-sensitive products. The back layer is translucent and gives creamish look and the middle layer is visible from back. FBBs are commonly used for packaging heavy weight as they are highly stiff quality wise. Grammage of FBB is commonly available in the range of 150-400 GSM.

3.4. Characteristics of Paper and Paperboard

Paper and paperboard have distinguishable characteristics compared to other packaging forms. Different international and national agencies have developed several standards to standardize the properties of paper material. British Standards (UK), DIN (Germany), ISO, TAPPI (Technical Association for the Pulp and Paper Industry), ASTM International (also known as American Society for Testing Materials) are some of the international agencies that are developing standards in different states of the world. Paper and paperboard satisfy packaging requirements as they offer aesthetic and functional qualities that make it possible to create a wide variety of package structures at a reasonable cost. Paper and paperboard have a flexible application as they can be used where the requirement is for flexible packaging as well as rigid packaging by cutting, creasing, folding, forming, winding, and gluing. These types of packaging have the property such that they are printable and varnishable and can be used to laminate materials on other materials also. Depending on the requirement of the product to be packed, the type of paper and paperboard are selected. Different types of paper and paperboard can be differentiated based on appearance and strength. These properties depend on the type of raw material used and the processing method. The characterization of paper is done based on its appearance, strength which depends on the type and amount of fiber used, and pulping methods, fiber treatment. In addition, the type of paper or paperboard material is also characterized by its weight per unit area and thickness.

The appearance of paper is measured in terms of whiteness, brightness, shade, texture and color, representing the quality of the paper. Weight per unit area is commonly used to describe the strength of paper. ISO defined weight per unit area as the characteristics dividing the line between paper and paperboard, which states that paper over 225 GSM will be considered as paperboard, board or cardboard. However, some products with lower grammage are also considered paperboard due to the improvement in manufacturing techniques, i.e. lightweight materials can now be produced with similar strength to older heavyweight grades. Different testing instrument are summarized in Figure 8 and different testing methods used for characterization of paper and paper board are discussed below.

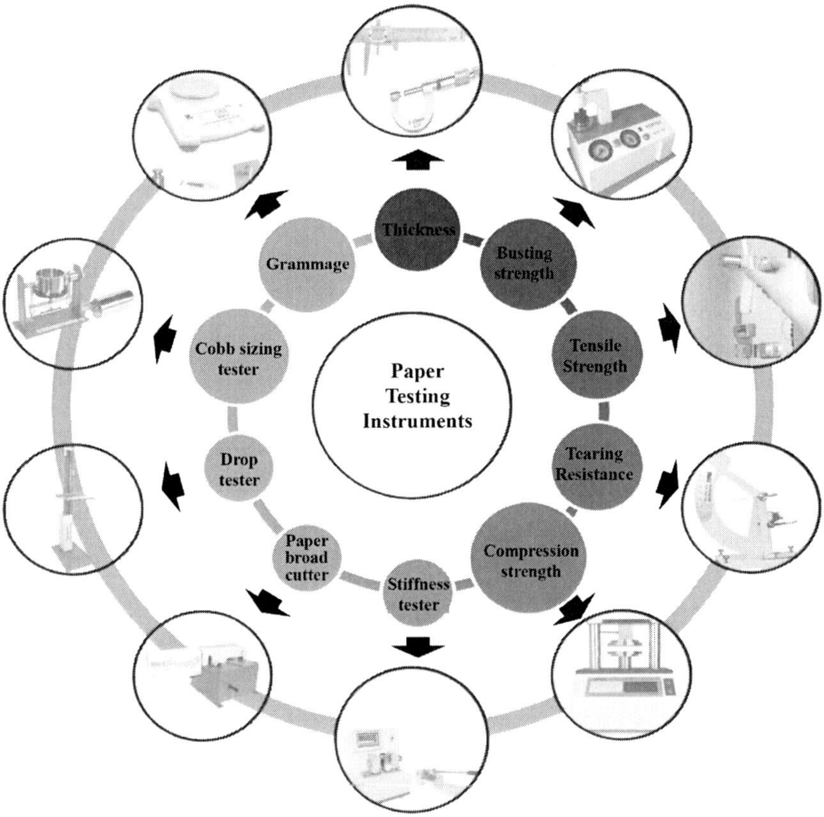

Fig. 8: Paper Testing Instruments

(i) ***Grammage (substance or basis weight):*** Weight is the most fundamental property of paper selection and has a profound impact on its application. It is also used as a one of the characterization criteria and commonly denoted as weight in grams per metre square (g/m^{-2}) of paper or grammage or GSM. It is also presented in pounds per 1,000 square feet (lb/MSF) or pounds per 3,000 square feet (lb/3MSF).

(ii) ***Thickness (caliper):*** Caliper is basically the thickness of paper or paperboard and is positively related to the strength and barrier properties of packaging material. It can be measured with instruments like a vernier caliper, screw gauge or micrometer. It is also measured by placing a paper sheet under a steady pressure of 0.49-0.63 kg/cm^2 between two circular parallel plane surfaces with an area of 1.6 cm^2 (0.25 $inch^2$).

(iii) ***Moisture content:*** The moisture content of paper is defined as the amount of water present in the paper. Moisture changes the properties of paper as it changes the composition of the fiber and has direct impact on the strength of the paper. Cellulose fiber is the main component of paper, and it absorbs moisture with increase in relative humidity (RH) and loses the moisture in low RH. This hygroscopic nature indicates that paper needs to be stored and transported with adequate care and coated with water-resistant material to avoid the damage.

(iv) ***Tensile strength:*** Tensile strength is the measure of rupture ability of a material which is the internal strength of the paper material. When force is applied perpendicular to the material surface, elastic deformation occurs upto a certain limit followed by rupture. It is normally measured by Testers/texture analyzer, wherein the paper is pulled apart perpendicular to the surface. It is an important consideration in multilayered paper and paperboard products.

(v) ***Stretch or elongation:*** Stretch is the maximum strain that can be produced before a material is ruptured in the tensile test under prescribed conditions. It is commonly expressed in percentage, i.e. cent times the ratio of change in length to the original length of the material.

(vi) ***Tearing resistance:*** Tearing resistance is defined as the resistance force against the first tear in the sheet. It is an important properties of paper and paperboard. It depends on the degree of fiber refining, inter fiber bonding, fiber length and the quality and quantity of filler used. Longer fiber increases tear strength as it distribute the applied stress over more fibers. Tear strength is more in lateral direction than in longitudinal direction. Insufficient or extra refining reduces tearing strength. Use of more filler also reduces tear strength. An Elmendorf type tester is used to measure tearing resistance where result is expressed as mN/ply.

(vii) ***Burst resistance:*** Burst resistance is the property of paper, indicating how tough the material is. It is measured by clamping a sheet of paper on a circular area and then increasing pressure on it till the time it gets ruptures. It can be increased by adding urea and melamine formaldehyde resin at the stock preparation stage.

(viii) ***Compression strength:*** Compression strength is the maximum compression force per unit width that a paperboard can withstand without buckling or bending. It is expressed as kN/m. Compression strength is important in storage, distribution and transportation, for packaging materials like cartons, cases and drums. Compression strength is different for the same material also, depending on the type of load. It can be calculated in dynamic or static load condition.

(ix) ***Stiffness:*** Stiffness indicates the resistance applied during the bending of the paperboard. Stiffness measuring device measures the force required to bend a strip of paper through certain angle. Some equipment measures the length on which a given force must act to produce a given deflection. It is defined as force, moment, angle or length that will produce a given effect on specimen of given dimension strained in specified manner. Taber stiffness tester is used to measure stiffness. It measures the force in Newtons to deflect a rectangular test piece, clamped at one end. through specified angle when the force is applied near the free end of the test piece, normal to the plane which includes the near edge if the test piece, the clamp and the point of line of application of force. This found major application in conversion, printing and other usages. Another term, stiffness ratio, is also used in the same context which is the ratio of machine direction (MD) and cross direction(CD).

(x) ***Creasability and foldability:*** Creasability and foldability are the two interlinked properties of paper. These properties have major applications in box or carton manufacturing. When cartons or boxes are manufactured, paper and paperboard are frequently folded to give them the proper shape. Foldability is the number of times a sheet of paper can be folded without any damage to the material. For thinner materials, folding is done mechanically through $180^{o,}$ and then the resulting folds are rolled to give permanence while for thicker materials, a crease is required to be made before folding. For creasing, the material is put on a thin sheet of paper known as make-ready (counter die) to give the packaging a proper shape.

(xi) ***Porosity:*** Porosity is defined as the ratio of the volume of pores or voids present in the matrix to the volume of solids in the matrix. The porosity of paper has a direct impact on the diffusion rate of substrate.

(xii) ***Water absorbency:*** Water absorbency can be defined as the amount of water absorbed by the wetted surface of paper and paper board materials. It is measured using the Cobb Test, in which a sized (non-bibulous) paper is kept in water for some time (60 or 180 sec) and water absorbency is measured based on the amount of water absorbed by the material. Water absorbency of material is useful in printability as well as in setting the rate of water-based adhesives. It is expressed in g/m^2.

(xiii) ***Gluability/Adhesion/Sealing:*** Gluability denotes the speed and strength of the bond created between the applied adhesive and the surface of paper. Gluability has a wide application when the shape of the package is of considerable importance. In the paper industry, rigorous testing protocols are applied to maintain the strength and quality of gluability of the package. Cold dextrin and hot melt glue tests are commonly used to find the long lastbility of adhesion in extreme conditions.

(xiv) ***Taint and odour neutrality:*** Odour in paper and paper board may arise due to several reasons which may be sometimes unpleasant and can deteriorate the quality food material to be packed. Odour can be due to wood resins from mechanical pulp or residual chemicals from chemical pulps. It may rise because of biological activity in the pulp or paper-producing substances. Sometimes, the coating used in paper manufacturing contains synthetic binders which also impart an odour to the paperboard.

3.5. Application of Paper-based Packaging

Paper has many advantages, including its environment-friendly tag that makes it a preference for many customers (Coelho et al., 2020). Paper-based packaging is currently observed in every industry, from juice or beverage to frozen food, where it is challenging to pack chilled foods. Paper based packages have found in all form of packaging as primary, secondary, and shipping (tertiary) containers (Figure 9).

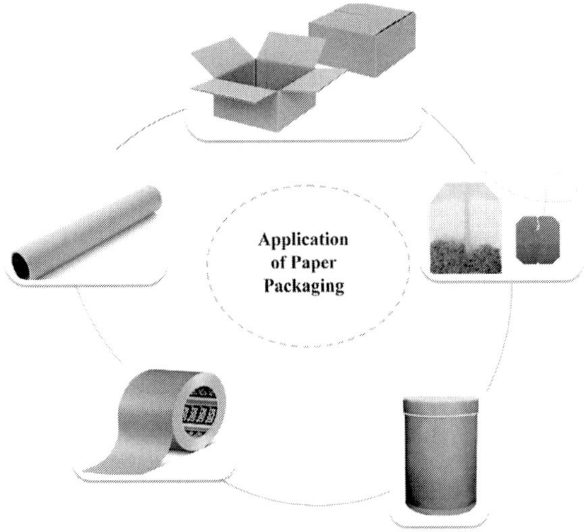

Fig. 9: Application of paper and paper-based Packaging in food Industry

3.5.1. Paper Based Packages

(i) ***Tea and coffee bags:*** Tea and coffee bags are made of very lightweight porous tissues using fully automatic process. These bags have high heat resistant properties as these don't even damage at 100 ° C of temperature. Heat resistance is provided by the materials such as PP. Lightweight (grammage ~17 g/m^2) material allows more infusion and leaching of material. These are available in different shapes and sizes like flat, square, four side perimeter sealed. Also sometimes, they are also round or pyramidal in shape. All bags are manufactured in the machine, which forms, fills and seals the bag and also attaches strings and tags with it. With the increase in technology, several types of bags are available that use highly lightweight material (~12 g/m^2) for packing.

(ii) ***Paper bags and wrapping paper:*** These are the most common type of packing material available in the market. Paper bags or wrapping paper packs do not need a packing machine and is the easiest form of packing. It is mostly used in retail shops, grocery shops, local markets, and bakeries. It is reusable form of packaging. Customers and retailers can also do manual wrapping of precut sheets for food products like fish, chips, popcorn and many more.

(iii) ***Sachets/pouches/overwraps:*** Sachets are the bag-like structure formed around the mandrels and sealing the top with adhesives. It can be distinguished based on the bottom of the sachet. A sachet is the safest option in terms of its barrier properties. It is a flexible form of packaging which involves paper with some other material like PE or plastic with layers of aluminum foil or metallised PET. It may be rectangular, or it can be a block shaped with square edges. It is used as packaging for dry product as well as for liquid food products. The base material for these sachets can be paper and plastic, more often PE. The inside aluminum layer provide excellent barrier properties against oxygen, light and moisture. Paper is used in the outer layer which allows easy tearing of the sachet. More layers can be incorporated into the material to provide special effects in packaging. Sealing layers like PE and ionomer-based can be used to suit different requirements of the customers. It also includes some added layers for heat-sealable products which do not let the product cool, whereas some cold sealable packaging is also available for heat-sensitive material. Packaging of material in sachets or pouches requires special machinery for packaging of food products. Form-fill and seal machinery are the most common way of packaging in sachets.

(iv) ***Multiwall paper sacks:*** Multiwall paper sacks are sacks made of two to six layers of paper or plies. These are designed based on the need of food product to be packed and consumer feasibility. The design of the opening, opposite endclosure, and side section are considered while designing the sack. Generally, brown kraft paper is used as it has high tensile strength with more percentage stretch, tensile energy absorption, burst strength, tear strength and where necessary, wet strength. These types of packages are commonly used when packing is to be done in bulk; however, small multiwall sacks are also available in the market. Hygroscopic food products such as sugar, dried milk, whey powder, coffee beans, flour, peanuts, potatoes and other fresh vegetable products are preferred to be packed in multiwall sacks. Multiwall sacks with layers of plastic film, foil, or polyethylene-coated paper are also used for water-repellent, insect-resistant, or rodent barrier.

(v) ***Folding cartons:*** Folding cartons are the most common and convenient forms of packaging for dry food products. These are also used for transportation purposes of various materials. It is the simplest form of packaging as a simple sheet of paper can be used just by folding them in many forms. Solid bleached and unbleached boards are the common material used for folding type cartons. Folded type cartons can be modified by lamination and coating with water-resistant materials. Dispersion and plastic extrusion coating can also be used for a special product type. These are available in many sizes, from small boxes of cookies or pastries to big cartons for transportation and storage of bulk material. The product to be packed can be in direct contact or already packed in some other form of packaging. It is used as a secondary form of packaging for a variety of food products, mainly for retail purposes. Cereals, frozen or chilled products, and sugar confectionery always require folded cartons to pack them securely. It is also used for packaging bottles, cans, sachet, collapsible tubes, plastic trays or pots. Folding type cartons are designed based on the product and its purpose of packaging (packing or distribution or storage).

(vi) ***Liquid packaging cartons:*** Liquid packaging cartons are usually made up of nearly 75% paperboard, 20% plastic, and 5% aluminum and its composition depend upon the layers of packaging used for the manufacturing of aseptic cartons. Refrigerated cartons consist of 80% paperboard and 20% polyethylene.

(vii) ***Rigid cartons or boxes:*** Rigid boxes are the most common types of packaging material used for bakery and confectionaries products. These are generally made of paper and paperboard material with thicknesses in the range of 1000 to 2500 microns and grammage of 400–1600 g/m^2. Several mechanisms are available in rigid boxes, out of which lift-off-lid mechanisms are the most commonly used. Rigid boxes are available in various shapes square, rectangular, round, elliptical etc., and designs such as shell and slide (drawer), book-style, flip-top, tray and lift-off lid. Rigid boxes are also common in the Indian market for the packaging of Indian dairy products too.

(viii) ***Paper-based tubes, tubs and composite containers***

Tubes are long hollow cylinder with small diameter used for packaging of confectionery and provide long lasting cushioning effect to the food product. It may be open or closed end with paperboard or plastic ends.

Paper based tubs are available in various shapes such as circular, elliptical or square with rounded corners and straight or tapered sides. The size of tubs is decided based on the amount of single-portion cream or yoghurt needs to be packed. Ice-cream tubs are leak proof containers which are highly resistant to the product and are designed for low-temperature storage. These are wax-coated paperboard or PE extruded to provide leak proof surface.

Composite containers are rigid structures drawn linearly with spiral or convolute bound with either one end open or both ends closed. These containers are available in different shapes, but the most common are round shaped. Oval and rectangular shapes are also available to attract the consumer attention. Kraft paper is used as base material for composite container with layers of polypropylene, high-density polyethylene (HDPE) and aluminum on the surface. Composite containers are considered as a major replacement for metal containers usually used to pack products like refrigerated dough, snacks, frozen fruits, dried fruits, nuts, chips, powdered foods, dried meats, chips, salt/spices, cookies/crackers, solid/liquid shortening and confectionery products which requires detachment from the environment and no physical harm on the food product. Different types of composite containers are describe below.

- ***Fiber drums:*** Fiber drums are big size cylinders for storage of food materials such as dry powders, chips, biscuits, cookies, pastes and semi-liquid food products. These are made of thick paperboard-based sidewalls with metal, plywood or paperboard as end components. They are commonly used for bulk storage of food material as side walls offer high strength and high protection to food products during transportation.

- ***Corrugated fiberboard packaging:*** Corrugated fiberboard packaging consist of layers of flat kraft paper (liner) with layers of corrugated material or flute corrugated sheets sandwiched between the flat layers to provides cushioning effect and abrasion resistance to packed material. It is produced using a corrugator in which flat kraft paper is passed between a series of serrated rollers to apply desirable pressure and to create adhesiveness inside the material. It mainly uses kraft paper and sometimes agave bagasse (a by-product from the tequila industry) is also used as a raw material. These types of materials are normally used to pack fruits and vegetables in direct food contact surfaces that have all grades of waste paper as internal layers. The specified requirement for the production of corrugated boards is the level of pentachlorophenol (PCP), phthalate, and benzophenone.
- ***Moulded pulp containers:*** Moulded pulp containers are the common containers used to serve delicate food products. These containers are produced from moulded pulp containing 96% water with only 4% fiber using water repellent substances such as wax and resin. Moulded pulp containers is made by molding its pulp into various shapes by pressing at high temperature and pressure and then drying the solution. The most common examples of moulded pulp containers are egg trays, fruits and vegetable trays which need a cushion effect in the product to preserve the delicate texture of food products. Recently, moulded pulp containers made of 80% wheat straw have been available in the market to provide better tensile properties and biodegradability.

(ix) ***Labels:*** Labels are identification characteristics in a packed food product. It contains nutritional information and guidelines, health claims, barcode, manufacturer's information, expiry date, manufacturing date, the product's value, amount of product, company's information. Labels sticks on the food packets using adhesives and sticking tapes. However, wet glue, gummed and self-adhesive paper labels are some new technology labeling which contains adhesion in labels only.

3.5.2. Paperboard Based Products

(i) *Folding Cartons*

Folding cartons are the most common application of paperboard having a thickness in the range of 300 to 1100 μm made by folding paper into the desired shape. The raw material for folding cartons can be kraft paper or recycled fiber from various sources. The outer layer in folding cartons can be made using white-lined chipboards, which gives an attractive look to the folding cartons. Bleached chemical pulps are also used to make different colored paperboard.

Commonly available paperboards are as follows

(a) *Coated Solid Bleached Board (SBB) or Solid Bleached Sulfate (SBS):*This grade of paperboard uses pure bleached chemical pulp with different layers of bleached coating on the top surface. The color after bleaching is chosen according to the food product and to increase customer attractiveness. It is commonly used for packaging chocolate confectionary, cheese, coffee, tea, reheatable products and frozen foods. This type of paper board can be used in combination with other unbleached boards to make it more cost-effective.

(b) *Coated Solid Unbleached Board (SUB) or Solid Unbleached Sulfate (SUS):*The coated unbleached board is the 100% virgin, chemically furnished board made of different layers of kraft paper with a variety of mineral coating on the surface. As it is made of coated material on kraft paper, it is also known as coated unbleached kraft (CUK). Unbleached Kraft pulp is sometimes replaced with recycled fibers. Nowadays, kaolin clay is also used to coat the unbleached board to improve its printability.

(c) *Folding Box Board (FBB):* This grade of paperboard typically consist of mechanical pulp sandwiched between two layers of chemical pulp with up to three layers of coating on the top or printing surface and one layer of coating on the reverse. The mechanical pulp provides stiffness and bulkiness while the chemical pulp imparts strength. It is used for products such as drinks, confectionery, frozen and chilled foods, tea, coffee, bakery products and cookies/biscuits.

(d) *White Lined Chipboard (WLC):*This grade of paperboard basically uses recovered fibers as raw material. It is manufactured in several layers, each of which uses selected grades of raw materials. It typically has two or three layers of coating on the top or printing surface, consisting of bleached chemical pulp. The second layer is made of bleached chemical or mechanical pulp, while the middle layers (commonly three in total) are unbleached recycled pulp from mixed paper or carton waste. The bottom or inside layer is specially selected recycled pulp (underliner). Some uncoated paperboard is produced with a top ply of white recovered fiber. WLC is used in a range of applications, such as frozen and chilled foods, dried food and breakfast cereals. In the absence of a functional barrier, mineral oil hydrocarbons (MOHs) from printing inks and recycled fibers tend to migrate from paper-based food packaging materials through the gas phase into dry food. The main sources of mineral oil in paperboard are offset printing inks, either directly applied for decorating the food packaging material or entering via recycling of fibers contaminated by mineral oil-containing inks, primarily from the newspaper.

(e) *Beverage Cartons:* The carton normally consists of layers of SBB and SUB coated internally and externally with LDPE, resulting in a carton which is impermeable to liquids and in which the internal and external surfaces may be heat sealed. There may also be a thin layer of aluminum foil, which acts as a gas and light barrier. The modern gable top carton retains the simple basic geometry of earlier years, although flat-topped and plastic-topped versions are available.

3.6. Waste Generation and Recycling of Paper

Paper production is an ancient process and the industry associated with it occupies third position with reference to water consumption. Paper production involves usage of large amount of water in processes like pulping and at the same time large amount of wastewater is produced. Not only water but also large amount of energy is consumed during pulping. As per the data, paper industry ranks fifth in terms of energy consumption and third in water consumption during manufacturing process. Indian paper industry which accounts for 3% of global paper production produces 13 million tons of paper every year. In the year 2014-15, 13.9 million tons of paper was produced which may increase upto 20 million tons by 2020 (FAO, 2021). Paper making process produces huge amount of land and water pollution with major pollution in the form of waste and wastewater generation, toxic emission that has deep impact on environment like exhaustion of natural resources, global temperature rise, ozone depletion and many more. In the paper producing industry, a lot of waste is generated which directly or indirectly affects the economics of paper mills as well raises issues related to environment. In addition to that, the forest harvesting for the production of paper is directly contributing to global-warming.

Recycling of paper has proved to be a best alternative for the reducing energy and water consumption and pollution. Recycling of paper requires less energy as there is no need of refining of paper from pulp. Studies have shown that there is less adverse environmental impact with low waste generation of recycled paper production when compared to virgin paper production. It has also been observed that there is less use of chemicals in recycling of paper which reduces pollution problem.

References

Chamberlain, D., & Kirwan, M. J. (2012). Paper and paperboard – raw materials, processing and properties. Handbook of Paper and Paperboard Packaging Technology: Second Edition, 1–49. https://doi.org/10.1002/9781118470930.CH1

Coelho, P. M., Corona, B., ten Klooster, R., & Worrell, E. (2020). Sustainability of reusable packaging–Current situation and trends. Resources, Conservation & Recycling: X, 6, 100037. https://doi.org/10.1016/J.RCRX.2020.100037

FAO publications catalogue 2021. (2021). FAO Publications Catalogue 2021. https://doi.org/10.4060/CB4402EN Forest product statistics. (n.d.). Retrieved June 1, 2022, from https://www.fao.org/forestry/statistics/80571/en/

Netramai, S., Kijchavengkul, T., & Kittipinyovath, P. (2016). Pulp and Paper Production. Reference Module in Food Science. https://doi.org/10.1016/B978-0-08-100596-5.03201-7

Smook, G. A. (2015). Handbook for pulp and paper technologists smook pdf.

Twede, D. (2005). The Origins of Paper Based Packaging. Proceedings of the Conference on Historical Analysis and Research in Marketing, 12, 288–300. https://ojs.library.carleton.ca/index.php/pcharm/article/view/1636

Ververis, C., Georghiou, K., Christodoulakis, N., Santas, P., & Santas, R. (2004). Fiber dimensions, lignin and cellulose content of various plant materials and their suitability for paper production. Industrial Crops and Products, 19(3), 245–254. https://doi.org/10.1016/J.INDCROP.2003.10.006

4

Metals in Food Packaging

Dhiraj Kumar Yadav and Vinkel Kumar Arora

Department of Food Engineering, National Institute of Food Technology Entrepreneurship and Management, Kundli, Sonipat, Haryana, India

4.1. Introduction

Packaging of materials protects, surrounds, and contains food products. It is necessary from processing and manufacturing stage till the final consumption through storage and handling. In 1804, Nicolas Appert, a French confectioner, discovered a method of "conserving all kinds of food substances in containers." In 1810, the Ministry of the Interior's Bureau of Arts and Manufactures paid Appert an ex-gratia payment of 12,000 francs on the condition that he publish the details of his process. Appert avoided the use of tinplate in his study. However, in the early nineteenth century (1810), Phillipe de Girard, a French inventor, filed a patent referring to the replacement of glass jars and bottles with tin cases. The official confirmation in the New York Evening Post on July 18, 1822 by Daggett and Kensett is thought to be America's first offering of preserved provisions in tin cans (Bishop, 1978). Table 1 shows the evolution of metal packaging materials at various stages.

Table 1: Progress in metal-based packaging containers

Breakthrough in metal packaging	**Year**
Nicholas Appert invented a method for storing food in containers	1804
Introduction of tin cans (Thomas Kensett and Ezra Daggett)	1822
Patented double seaming can ends by Max Ams of New York	1896
Aluminum foil production in USA	1913
Aluminum foil lamination on paperboard	1921
Introduction of 3-piece tinplate can	1935
Introduction of two-piece aluminum can and production of retort pouches	1950
Initial production of the first aluminum drawn and wall ironed (DWI) cans by a US brewery	1958
Japan pioneered the development of ECCS	1960
Introduction of tinplate can	1970
Application of aluminum bottle can for energy drinks	2001

Currently, out of the total packaging materials required in the globe, 15% of packaging materials are used as metal (Deshwal & Panjagari, 2020). The beverages industry leads a market share of 75% in the food products industry. A market survey predicts that the metal cans market will reach US $ 61.37 billion by 2026, with a cumulative annual growth rate (CAGR) of 2.23 % from its current value of US $53.37 billion (Mordor Intelligence, 2021). Countries like North America and Europe leads the metal market share.

For food packaging applications, various metals such as aluminum, tinplate, tin free steel (TFS) and stainless steel are used. The most typical forms of metal-based packaging include rigid and semi-rigid cans, foil wraps and retort pouches.

4.1.1. Performance of Metal Container

To deliver the food products to the final user i.e. consumer in a safe and hygienic manner the metal container must have to perform the following basic functions:

(i) Contain and protect the product.

(ii) Able to resist external environmental conditions.

(iii) Should not react chemically with the food product.

(iv) Printing capability for leveling (container must be leveled with sufficient information) and easy to open.

4.2. Needs of Metal Container in Food Packaging

Because of their magnetic properties, metal packaging materials have a lower global warming potential and higher recyclability, allowing for easier segregation. Metal-based packaging materials have excellent barrier properties and are thus commonly used in food packaging applications. Product safety, hygienic, resistance to transportation, air tight packaging and easy recyclability are the significant drivers of the metal market. When compared to raw materials, manufacturing products made from recycled metals, use less energy, emits less carbon dioxide (CO_2), and use less water. Another important need of metal container is increasing alcoholic and non-alcoholic beverage consumption such as cold drink i.e., carbonated drink, beer and wine that supports the evolution of metal can market around the world.

4.3. Advantages and Disadvantages of Different Metal Packaging

Metals containers like tinplate, aluminum, steel, chromium containers and closures are universally adopted for food and beverages.

Advantages of metal containers

(i) High mechanical strength and resistance to wear and tear.

(ii) Low toxicity of materials.

(iii) Ability to withstand wide extremes of temperature.

(iv) Excellent resistance to moisture, light, and vapors.

(v) The metal surface lends itself well to lacquering and coating.

(vi) Hermetically sealed cover.

Disadvantages

(i) Some metals are corrosive in nature which can react with the food products and influence their quality.

(ii) It is not possible to see the food product inside the container as it is possible in glass packaging.

(iii) It is moderately heavy than other packaging materials.

(iv) Metal manufacturing is a multi-stage process hence it takes longer time and is not very cost effective.

(v) Metals are not flexible as paper or polyethylene-based packaging due to this reason sometimes it creates problem during storage because it cannot be bend.

4.4. Types of Metals used in Food Packaging

Four major types of metals are used in food packaging which are given below

(i) Aluminum

(ii) Steel

(iii) Tin

(iv) Chromium

The usage of metal container can be broadly classified into two categories.

(i) General line container

(ii) Open top sanitary (OTS) can

General line containers range from a tiny line container of the size of a biscuit to the large 15 or 20-liter edible oil can. These are available in round or variable other shape and fitted with various kind of fitments made from aluminum, tinplate and plastics. The general line container is used for various purposes such as edible oil, confectionary, aerosols, talcum powder and paints etc. However, it is not used for processed food packaging. General line container can be constructed in two ways. First is a built-up container which has a body, top and bottom that are assembled with each other by seaming. The body itself has a side seam because it has been rolled from a flat blank and hence its folded edges need to be engaged and pressed. Another type of general line container is seamless container which has no seam at all. The seamless container makes it possible to manufacture containers with small heights which would be very difficult on built up type. Ratio between height and diameter is very important in feasibility and success of seamless container (Board, 2010). Because of small height of the seamless container, it requires less manpower and equipment and lead to greater integrity of the container.

OTS cans are used for packing of food products such as milk powder or any other hygroscopic or processed products. This type of cans is also called as sanitary cans and essentially round in shape. These cans have to undergo a process of either pasteurization, sterilization and then go for vacuum packing or nitrogen flushing for getting a prolonged shelf life in a hygienic manner. Application of OTS can in pharmaceuticals are packing of tablets, powder, and capsules as well as also used for trade packs. OTS container can be used for processed food which is not packed in general line container. The open top sanitary can is subdivided into two types.

a. Two-piece can

b. Three-piece can.

Two pieces cans are generally manufactured from Aluminum by drawing, redrawing and ironing process. Two-piece cans consists of one end (lid) and a seamless body without any joint. It does not have side seam like three-piece cans. For this reason, two piece cans are hygienic, economic, larger printing area and approximately 35% lighter than three piece cans (Deshwal & Panjagari, 2020). Manufacturing of two piece can involves two major processes: (1) drawn and redrawn (DRD) and, (2) drawn and wall ironed (DWI).

Drawn and redrawn (DRD)

DRD cans are thicker and can endure the headspace vacuum created during cooling following heat sterilization or retorting. DRD manufacturing is similar to the initial stages of the DWI process, except that instead of ironing to reduce layer thickness, metal is moved from the base of the container to the wall by reducing the diameter of the container. Internally, epoxy, phenolic, or vinyl-based lacquers are used for both DWI and DRD processes to prevent metal-to-product interaction.

Drawn and wall ironed (DWI)

DWI process produces thinner walls than the DRD processes and is used to make aluminum cans for carbonated drinks where the gas pressure supports the containers. A disc-shaped plate is cut and stretched into a cap during the DWI manufacturing process. This is forced through a series of annular rings (wall ironed) to decrease the thickness of the metal wall and raise the can to the appropriate height.The base thickness of a DWI can is greater than the wall thickness because only the side wall is stretched and thinned during the ironing process.

Three-piece can is generally manufactured from tinplate and depicted in figure 1. Tinplate cans can be supplied in flattened condition to reduce the transportation cost. Aluminum can are supplied always in buildup condition (formed body from single piece of metal). Tinplate cans can also be supplied in buildup condition (formed body with attached one lid). Three pieces can formation involves two end lids and a body. To initiate, cut metal sheets are coated, printed, and rolled as needed, and the longitudinal sides of the flat metal sheet are joined either by soldering or welding. Welding is preferred, but soldering is usually avoided for food cans due to concerns about lead migration from tin/lead solder into foods. Welding eliminates the safety concerns associated with soldering while also reducing metal consumption. To interlock the metals, soldering requires more metal as compared to welding (Deshwal & Panjagri, 2020). After the formation of cylindrical body by welding or soldering, two important processes are performed i.e. 1) necking and flanging for beverage cans, and 2) beading and flanging for food cans.

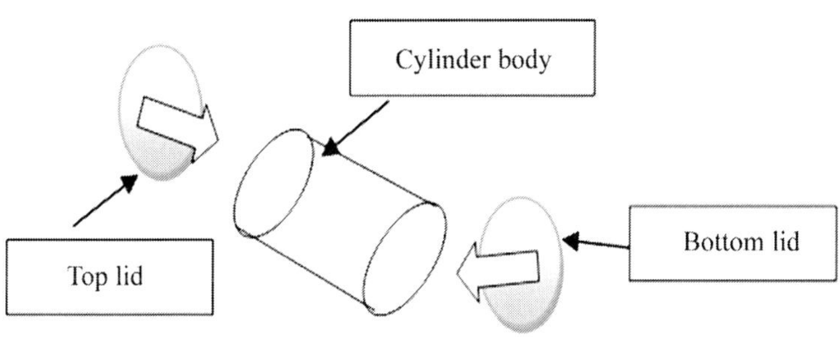

Fig. 1: Three-piece can

4.5 Major Types of Metals used in Food Packaging

Four major types of metals are used in food packaging which are given below

(i) Aluminum (ii) Steel (iii) Tin (iv) Chromium

Composite metals-based food packaging materials:

(a) Aluminum alloyed container

(b) Tinplate container (SS + Tin coated)

(c) Electrolytic coated Chromium container (Tin free steel containers)

4.5.1. Aluminum in Food Packaging

Aluminum is widely used due to its light weight and excellent malleability. Aluminum is used in relatively pure form for light metal packaging, with magnesium and manganese added to increase the strength. It cannot be welded by can-making systems and is therefore only suitable for seamless (two-piece) containers. Because of the products that are typically packed, the insides of aluminum containers are always lacquered with an organic lacquer. However, aluminum is used in pressurized containers where strength of wall is not as important, for products such as beer and carbonated drinks in DWI cans, as well as seamless impact extruded aerosols and collapsible tubes. Roll-on pilfer proof caps can only be made of aluminum because the metal must be soft enough to conform to the shape of the underlying bottle thread. Most commercial applications of aluminum necessitate special properties that pure aluminum cannot provide. As a result, a diverse range of aluminum alloys with high strength, improved formability, and low corrosion is available commercially for packaging purposes.

The aluminum alloy is used to enhance its properties (mechanical, physical and chemical) and usefulness, however it depends on composition and microstructure. The alloying elements used in pure aluminum are manganese, magnesium, copper, silicon, zinc, etc. Broadly, aluminum alloy can be divided into two categorized such as wrought alloy and cast alloy. Based on phase solubility, many alloys respond to thermal treatment. Heat treatment, rapid cooling (quenching), and age hardening are examples of these treatments. Aluminum alloys are reinforced and strengthened through either room-temperature deformation known as strain hardening or aging heat treatment.

Here is a brief summary of the effects of the principal alloying elements used in aluminum. The emphasis is on their impact on strength and heat treatment response. One of the most important components of aluminum alloys is copper.

Copper has a significant solubility and strengthening effect due to the age-hardening effect it imparts to aluminum. Many alloys exhibit copper as a major addition or even as an additional alloying element in concentrations ranging from 1 to 10%. Magnesium significantly strengthens and improves the work-hardening attributes of aluminum. It is relatively soluble in solid aluminum, but Al-Mg alloys containing less than 7% Mg do not exhibit significant heat treatment characteristics. It can also be used in conjunction with some other alloying elements. Zinc is used in casting alloys and also in wrought alloys with magnesium to produce heat-treatable alloys with the highest strength among Al alloys. Furthermore, copper and silicon are frequently used in casting alloys. Heat treatable and non-heat-treatable alloys both have desirable ranges of properties and characteristics. Mg_2Si is formed by adding an appropriate proportion of Mg-Si as a basis for age hardening in both wrought and casting alloys. Tin improves aluminum's antifriction properties, and to decrease density and increase elastic modulus. Lithium is added to some alloys at levels close to 3 % (Davis, 1999). The chemical composition and typical usage of some of the more commonly used aluminum alloys (the aluminum is at least 99% pure) are shown in Table 2.

Table 2: Aluminum alloy composition (%weight) and typical application (Robertson, 2005)

Alloy	Si	Fe	Cu	Mn	Mg	Cr	Zn	Ti	Application
1050	0.25	0.4	0.05	0.05	0.05	0.05	0.03	0.03	Foils and flexible tubes
3104	0.60	0.7	0.25	1.4	1.3	-	0.25	0.10	Beverage can end and D&I can body
5042	0.20	0.35	0.15	0.5	4.0	0.10	0.25	0.10	Easy-open beverage can end and tabs
5182	0.20	0.35	0.15	0.5	5.0	0.10	0.25	0.10	Easy-open beverage can end and tab
8011	0.90	1.00	0.10	0.20	0.05	0.05	0.10	0.08	Pilfer-proof cans
8079	0.30	1.3	0.05	-	-	-	0.10	-	Foil for lamination

The following are the general effects of various alloying elements on the corrosion behaviour of aluminum:

(i) Copper minimizes aluminum's corrosion resistance more than any other alloying element, resulting in a higher rate of general corrosion.

(ii) Manganese slightly increases corrosion resistance.

(iii) Magnesium has a significant effect, and Al–Mg alloys are resistant to corrosion.

(iv) Zinc has a minor impact on corrosion resistance in most environments, tending to reduce alloy resistance to acidic media while increasing resistance to alkalis.

(v) Silicon reduces corrosion resistance slightly depending on its form and location in the alloy microstructure.

(vi) Usual amounts of chromium added to alloys, increases corrosion resistance.

(vii) Iron decreases corrosion resistance and is most likely the most common cause of pitting in aluminum alloys. A high iron content increases bursting strength but decreases corrosion resistance.

(viii) Titanium has little effect on aluminum alloy corrosion resistance.

a) Aluminum used as foil and container in food packaging

The term aluminum foil refers to 99 % pure aluminum that is available in the form of thin rolled flat sheets ranging in thickness from 4 to 150 microns. There are several aluminum foil alloys available, including 1100, 1124, 1235, and 3003. Alloy 1235 is commonly used in flexible food packaging applications, whereas alloy 3003 is used for thicker and stiffer aluminum. At a minimum thickness of 15 microns, it is impermeable to liquid and gas. The main step involved in aluminum foil production is described by flow chart in figure 2.

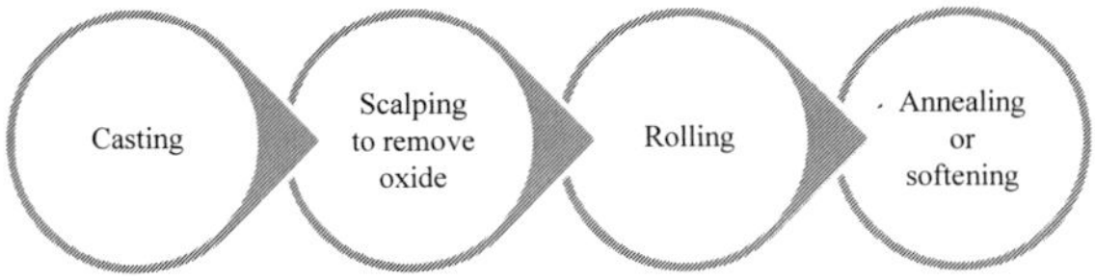

Fig. 2: Process flowchart of aluminum foil production

b) Aluminum foil in retort pouch

Retort pouch is made from one or more than one layer of foil or plastic. It is a flexible package, developed in the year of 1950s, sealed hermetically from three or four sides. The thin profile of the package (12–33 mm for 200–1000 g pouches) is one of the advantages of the retort pouch over the metal can, allowing retorting times to be reduced by up to 60%. A typical three-layer pouch structure may have an outer layer of 13 μm PET for strength and toughness, a middle layer of 9–18 μm aluminum foil for moisture, light, and gas barrier, and an inner layer of 70–100 μm CPP for heat sealability, strength, and food compatibility. If longer shelf life is required, 15–25 μm BON-6 can be added as an inner layer of pouch. The advantages include the ease of reheating and serving, saving in weight and space, ease of carrying with product safety and simpler in disposal than metal can.

Advantages of aluminum in food packaging

(i) Low cost as compared to steel and non-toxic material.

(ii) Easily available in the local market.

(iii) Can be considered as a sustainable packaging option because it is more frequently reused and recycled.

(vi) Due to the good recyclability of aluminum, there is minimal amount of pollution associated with it.

(v) It has highest strength-to-weight ratio of any metal.

(vi) Relatively corrosion free.

(vii) Good internal pressure resistance.

(viii) Good aesthetic appeal and bright & clean image in the market.

Limitations of aluminum in food packaging

(i) Aluminum cannot be used to make three-piece welded side seam containers.

(ii) Aluminum can/container must be handled carefully during product filling, transportation, storage and handling.

(iii) As it always requires coating with organic lacquers therefore, it is less economical for food stuff.

(iv) High cost machineries are required for product filling compare to tinplate containers.

(v) Skilled manpower requirement.

4.5.2. Steel in food packaging

Tinplate is used to make the majority of steel-based packaging because the presence of tin on the surfaces provides good resistance against corrosion. Based on their crystalline structure, the three major types of stainless steel are austenitic, ferritic, and martensitic. Table 3 depicts the mechanical and chemical properties of the three major types of stainless steel mentioned above. Austenitic types are food grade and are most commonly used in packaging applications. It is primarily used in food packaging for returnable containers (for wine, beer, and soft drinks). Stainless steel is the most popular material for storage or transportation containers. Due to the presence of 2% molybdenum in the latter, the food and dairy processing industries primarily use austenitic

304 grades (18% Cr and 8% Ni) for mild treatment applications and austenitic 316 grades (16% Cr and 10% Ni) for a highly corrosive surface (Cvetkovski 2012).

Advantages of steel in food packaging

(i) Easily available and non-toxic material.

(ii) It has no chemical effect and can have direct contact with food products.

(iii) Can be used to create any type of metal package except when impact extrusion is used to form a two-piece seamless body.

(vi) It provides a complete protection against heat, water and air.

(v) It can maintain the desired temperature and also easy to clean and recyclable.

Limitations of steel in food packaging

(i) It is not normally used for welding because the chromium surface is incompatible with the welding process.

(ii) It is more expensive than aluminum and tin cans.

4.5.3. Tinplate container in food packaging

It is widely used in food packaging and about 50% of tin production is focused on the tinplate manufacturing. Tinplate consists of carbon steel and thin coating of $FeSn_2$, thickness ranging from 0.13-0.40 mm (Sarkis et al., 2011). Demand for tinplate can has evolved due to its lighter weight, good strength, reusable, temperature tolerance, ease of recycling and consumer acceptability (Dey and Agrawal 2016). The tin in food packaging is mostly used as a coating material. Tin coating plays an important function in establishing shelf-life because it protects the structural components of the can i.e., steel base plate. Tinplate is used either as cans, containers or caps. In tinplate, phenomenon of corrosion is major drawback, to overcome this limitation the production of tin free steel (TFS) is achieving great importance in food industry (Jafari et al., 2018). Owing to the complexity of welding, TFS can only be used in the top and bottom. Taroca et al. (2020) investigated electrochemical corrosion in tomato paste tin-free steel.

Table 3: Mechanical and chemical properties of some common types of stainless steel

SS Types	JSL Grades	Mechanical properties				Chemical compositions						
		Tensile strength (MPa)	Yield strength (MPa)	% Elongation	Hardness Rockwell B	%C	%Mn	%P	%S	%Si	%Cr	%Ni
Austenitic	J-201	655	310	40	100	0.15	5.5-7.5	0.060	0. 030	1.0	16-18	3.50-5.50
	J-202	620	260	40	100	0.15	7.5-10	0.060	0.030	1.0	17-19	4-6
	J-4	650	325	40	100	0.10	8.5-10	0.080	0.010	0.75	16-18	4-6
	J-301	515	205	40	95	0.15	2.0	0.045	0.030	1	16-18	6-8
	J-304	515	205	40	92	0.07	2	0.045	0.030	0.75	17.50-19.50	8-10.50
	J-316	515	205	40	95	0.08	2.0	0.045	0.030	0.75	16-18	10-14
	J-317	515	205	35	95	0.08	2.0	0.045	0.030	0.75	18-20	11-15
Martensitic	J-410	450	205	20	96	0.08-0.15	1	0.045	0.030	1	11.50-13.50	0.75
	J-415	795	620	15	32HRC	0.05	0.50-1	0.030	0.030	0.6	11.50-14	3.50-5.50
	J-420	690	-	15	96	0.15	1	0.040	0.030	1	12-14	0.75
	J-405	415	170	20	88	0.08	1	0.040	0.030	1	11.50-14.50	0.60
Ferritic	J-430	450	205	22	89	0.12	1	0.040	0.030	1.0	16-18	0.75
	J-441	430	250	18	88	0.030	1	0.040	0.015	1	17.50-18.50	-
	J-444	415	275	20	96	0.025	1	0.40	0.030	1	17.50-19.5	1
	J-446	515	275	20	96	0.20	1.50	0.0040	0.30	1	23.0-27	0.75
Ferritic + Martensitic	J-409M	450	275	20	90	0.030	.8-1.5	0.030	0.030	1	10.50-12.5	1.5

The effects of pH, temperature, and NaCl were investigated. At low temperatures (below 22 °C) and NaCl concentrations ranging from 500 to 2000 ppm, pH has no effect on corrosion rate. Furthermore, for NaCl concentrations less than 2000 ppm, temperature appears to have no effect on corrosion rate. Working with a chloride concentration less than 2000 ppm, the corrosion rate appears to be constant regardless of pH or temperature. Commonly required materials for tinplate manufacturing are tinplate, lacquering compounds, lining compounds in the lid, copper wire for body welding, and Al foil for tagger.

Tinplate does not have complete inertness therefore; it is often protected with organic coating from inside and outside of the body. Grammage of the coating ranges from 3-9 g/m^2. Tinplate is created by rolling steel in a rolling mill, removing mill scale with acid, and then coating it with a thin layer of tin. Tin plating can be done by three different methods such as barrel plating, rack plating and vibratory plating. Each of these processes is based on the accumulation of an electrolyte solution onto the surface of a metal element.

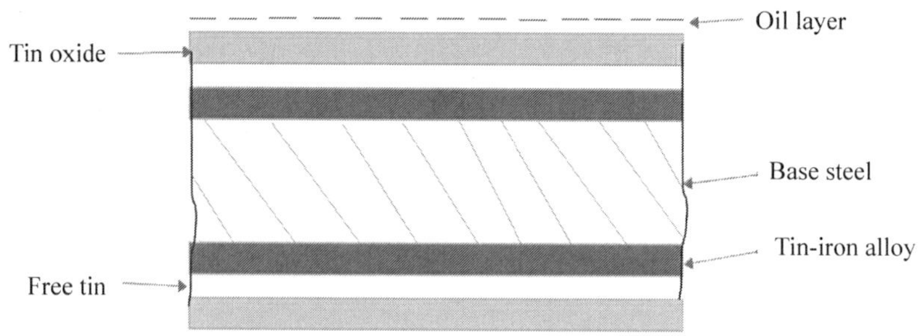

Fig. 3: Basic structure of tinplate container

The above tinplate cross section (Figure 3) depicts the layers present in tinplate. The tinplate consist of tin-iron alloy ($FeSn_2$) layer, base plate of stainless steel, free tin coating, tin oxide formed by passivation process and oil layer. On the surface of tin, a oxide layer is formed naturally and grows in the atmosphere. For the purpose of improving the stability and weather resistance of the steel strip's surface, passivation is applied that results as a thin layer (less than 1 μm) of film. After passivation, the plate is lightly oiled to protect it from corrosion and to make it easier for sheets to pass through container-forming machinery without destroying the thin layer of tin. The amount of oil is carefully regulated since a thick oil coating can cause the inconsistent application of printing inks and enamels during the subsequent container manufacturing process. The strips are then coiled or sheared into sheets and packaged for delivery to the can producers.

- Thickness of steel sheet – 0.15 mm to 0.5 mm
- SS plate Type L – High purity – High internal corrosion resistance
- SS plate Type MR – higher amount of copper and phosphorus and widely used in vegetable and meat.
- SS plate Type N – Nitrogenized steel – Higher strength – Can ends and aerosol domes
- SS plate Type D – Less carbon (0.13 decreased to 0.12%) – Non aging – Used for severe drawing operations
- Tin coating thickness – 0.4 μm to 2.5 μm (micron)
- Tin mass per unit area – 2.8 gsm to 17 gsm
- Differential tin coating – Two surfaces of tinplate having different tin coating
- Standard Designations for tin coating – D11.2/5.6, it indicates that 11.2 GSM of tin coating on one surface and 5.6 GSM on other side. Higher tin coating surface will go product side and lower one obviously to external side.

a) Manufacturing process

A process flowchart of manufacturing process is given below in figure 4. Tinplate is internally coated with the necessary lacquer (synthetic resin) and cut to size according to design specifications to fit the printing machine. Following the slitting of the printing sheets to the appropriate sizes for the blanks, the actual can-making process begins with a number of steps, including hooking the blank body, soldering and welding the side seam, flanging, and lastly seaming the ends. In printing process, a thin coating of lacquer is applied to sized sheets to provide decoration and base on the background of tinplate for printing and then cured in hot air oven at high temperature. The printing stage is completed by offset process using various color, depending on the needs. Rotary or scroll slitters are used to make sized tinplate into body blank. The vertical edge of body blank is folded to make outer and inner hooks.The flat body blank is formed into specific shape of the can either by rolling or rounding. The hooks are engaged and pressed together to form a side seam. Gaps (pin holes if any) in between side seams is filled by molten solder or by side seam cement to make it leak proof. Finally, can open ends are curled (flanging).

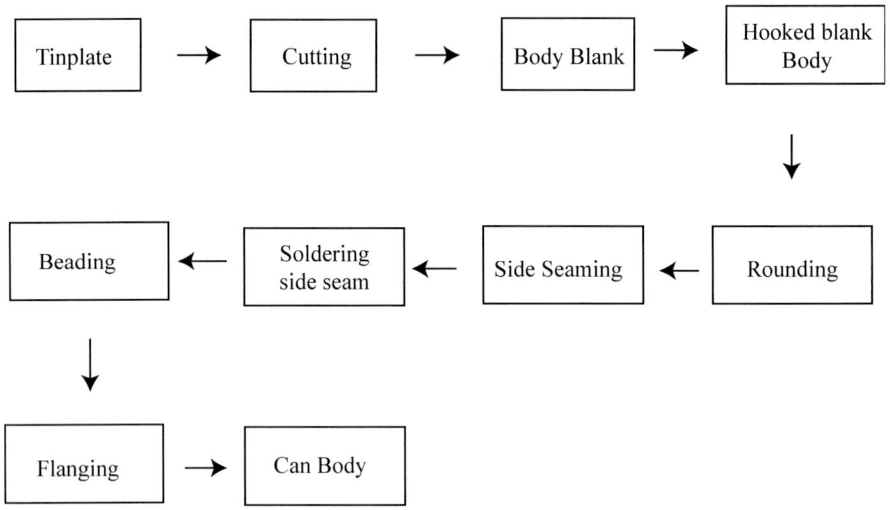

Fig. 4: Manufacturing process of tin can

b) Lacquering compounds

- To provide additional protection
- To provide an inert barrier between the metal of the container and its contents

c) Characteristics of lacquering

- Must not impart any odour or taint to the food and drink coming into contact with it.
- An ability to spread evenly,
- To adhere to the surfaces under what may be adverse conditions.
- For processed food, cans must be capable of sustaining one hour at temperatures in the range 120 - 132°C

d) Common lacquering compounds

Protective lacquers are used to cover metals to prevent migration and interaction with the food because metals do not react favorably with food products. The thickness of lacquering varies from 4 micron to 12 micron. Generally, lacquering mass per unit area varies from 3 GSM to 9 GSM. Epoxy Phenolic resins, Vinyl acrylic, Polybutadiene, Sulphur resistant lacquering compounds are the common lacquering compounds used in tin coating. Recently, the application of lacquers becoming common because of their tin dissolution. The most common types of lacquers used today are Epoxy phenolic group, vinyl resins, White vinyl, Polybutadiene, Sulphur resistant lacquering compounds, and others. Epoxy phenolic group are suitable for fruit, vegetables as well as meat and fish products. White vinyl lacquers have been used in situations where staining of the underlying metal due to product reaction is an issue and also used in hygienic/clinical appearance rather than the aesthetically undesirable corrosion patterns. Vinyl resins are ideal for dry packs because they are taste and odor free.

e) Tinplate corrosion mechanisms during product storage

Mechanical damage to cans, such as denting from improper handling, can cause internal lacquer cracking. This allows the product to obtain access to the underlying metal, which can result in quite rapid localized corrosion depending on the can and the product. Internal corrosion can occur on rare instances as a result of an incredibly aggressive reaction among the can and its contents, having caused the lacquer to detach from the can surface. The causes of these reactions are frequently complex, and occasionally changing lacquers is the only remedy.Under the improper conditions of storage, corrosion of the tin is more rapid than the normal condition.

f) Factors affecting tinplate corrosion

Time and temperature, exposure of the tinplate, tin-coating weight, type and composition of the product, presence of certain ions and vacuum level are the major factors which are responsible for the tinplate corrosion (Coles et al., 2003).

Tin corrosion is a continuous and time dependent process that is mostly influenced by storage temperature. Initial the rate of corrosion is higher in storage. Exposure of the tinplate influence the corrosion process. For example, containers with little or no exposed tin will have low tin dissolution, whereas products with even partial exposure will dissolve tin at elevated concentration. Tin crystal size, tin coating thickness, and passivation treatment all affect the rate of tin dissolution into the product. It has been found that thinner layer of tin coating increased the rate of dissolution than the thicker layer. Product compositions such as acidity, pH etc. have a direct effect on tin corrosion. Some chemical factors have also significant effect on rate of corrosion such as residual oxygen and nitrates (cathodic depolarizers). These chemical factors may arise from product ingredient itself such as process water, sugar, pesticide residue and food additives etc.

g) Precautionary measures to restrict tinplate corrosion

To prevent corrosion, practical ways such as incorporation of corrosion control in design process and protective corrosion barriers should be applied. Coatings can inhibit metal corrosion parts by providing a barrier between both the metal and oxidizing elements in the environment. Galvanization is a common method wherein the manufacturers apply a thin coating of zinc to the part. Corrosion can also be prevented by applying an opposing electrical current to the metal's surface. An impressed current is a type of cathodic protection that employs an external path of electrical current to overcome a corrosive current in the component. Some of the common precautionary measures used to restrict tinplate corrosion are listed below.

(i) Appropriate tin coating and lacquer selection.

(ii) Nature of product (pH and acidity).

(iii) Appropriate head space and vacuum in container.

(iv) Nitrite and nitrate content, SO_2 content of the product and the process water.

(v) Prompt cooling of cans after retorting.

(vi) Anthocyanins and other pigment content of food product.

(vii) Storage of cans at appropriate place (storage temperature and humidity).

(viii) Synthetic coloring compounds in product.

h) Advantages and limitations of tinplate containers

Advantages

(i) It is cheaper than aluminum.

(ii) It is recyclable.

(iii) It has a magnetic behavior that assist in its separation.

(iv) It is easy to print.

(v) It is impermeable to moisture and vapors.

(vi) It can withstand high temperatures during product processing, making it suitable for sterile products.

(viii) It can withstand internal pressure.

Disadvantages

(i) It necessitates surface coatings because it may react with food, and its containers generally necessitate the use of an opener to gain access to the product.

(ii) Few short-term symptoms like diarrhea, fever, nausea, abdominal bloating, vomiting and headache might occur because of affected gastrointestinal tract due to high concentration (200 mg/kg) of tin in food.

(iii) Flexibility in can shapes and sizes are very limited. Standard tin can sizes are only available.

4.5.4. Chromium in food packaging

Chromium is a highly active metal, but when it comes into contact with oxygen in the atmosphere, it forms an inert layer of chromium oxide (Cr_2O_3) on the steel surface, resulting in auto-passivation against corrosion. Chromium in food application mainly used as coating material in the form of electrolytic coated chromium container.

4.5.5. Electrolytic coated chromium containers (ECCC)

ECCC is also known as tin free steel (TFS) or electrolytically chromium/chromium oxide coated steel (ECCS). It is similar to a tinplate except that it is manufactured without the use of flow melting or chemical passivation. Dual electroplating of chromium and chromium sesquioxide is used in the production process, followed by coating with an oil such as butyl stearate oil. Electro chromium coated steel (ECCS) is widely used in the food industry to manufacture can ends and two-piece cans. Traditionally, they are manufactured with hexavalent chromium-based chromic acid electrolyte baths and composed of a low carbon, low gauge (0.13–0.49 mm) steel substrate covered in chromium metal and hydrated chromium (III) oxide layers. Substitute of the hexavalent chromium used for such baths is now a research trend due to health and environmental concerns (Wijenberg et al., 2015). One of the proposed substitute for packaging and decorative applications is trivalent chromium-based coatings (Del Painta et al., 2018; Leimbach et al. 2018). Chromium-based coatings are commonly used to provide resistance to corrosion to packaging steel substances, which are shown to corrode when exposed during the deformation process during can manufacturing (Williams, & McMurray, 2009).

It has been reported that a chromium and chromium (III) oxide coating combinatorial library can be created on a single substrate using physical vapour deposition (PVD) of a "wedge" onto a steel substrate. This enables a rapid, parallel research is being conducted to determine the role of coating thickness in attempting to prevent corrosion-driven organic coating from packaging steelusing scanning kelvin probe (SKP). The chromium coating thickness and the parabolic rate constant of cathodic disbandment have been found to have an almost linear, inverse relationship (Wint et al., 2020). Chromium coating has several advantages and disadvantages as given below.

a) Advantages and limitations

ECCS has several advantages over tinplate containers. Some of these advantages are given below:

(i) Due to the absence of a sacrificial tin layer, ECCS is less expensive and more susceptible to corrosion in an acidic environment.

(ii) Due to low melting point (232 °C), is more susceptible for protective enamel coating.

(iii) Due to the low cost of TFS, it is the best choice for the storage of bulk material and transportation of end products.

Limitations

(i) The requirement to remove coatings before welding TFS limits its widespread use for single-use containers and recyclability.

(ii) Because of the lacquer coating, it does not have a reflective surface like tinplate.

4.6. Metal Closures

The metal closures play very important role in sealing of beverages glass bottles in the form of pilfer proof. Examples are shown in figure 5.The crown cap (with a thickness of 0.25 mm) is typically made of tinplate, tin-free steel, or black plate. The crown cap's internal and external surfaces should be finished to provide protection and decoration, and they should be free of pit holes and off-flavors. Using hot melt type adhesive, line the inside surface of the crown cap is lined with cork or plastisol liner. Glass jar and bottle top lids are typically made of 20–70-micron aluminum foils coated with polyacrylic or polyvinyl chloride. For baby foods, jams, pickles, and dry powders, vacuum closures such as lug caps and twist-off type are used (Peter and Ulrich 2007). Re-closable cans are gaining popularities in the market due to its user-friendly mechanism. It is available in the range of 200 mL of 1.0 L (Dey and Agrawal 2016). Resealed cans are pressure stable against gas and liquid and works by pulling up an integrated opening/closing strap, then sliding it backward and forward for drinking and closing respectively.

Fig. 5: Metal closures

a) Important terminology related to metal packaging

Necking is the process of reducing the diameter of flat ends in a perfectly cylindrical seamed body to concave ends, thereby reducing metal usage.

Body: The rolled flat metal sheet with longitudinal sides joined to form a cylindrical structure with open top and bottom.

Beading: The beading treatment of food cans adds spherical curves (corrugations) on the round can body structure for added strength during retorting of food products. It facilitates handling of the can.

Flanging: The process of making outward flanges (hooks) to improve the fit of can lids (can ends).

Double seaming: It entails the interlocking of the flange and the edge of the lid.

Tinning is a common term for tin coating, which is formed through an electrolytic process associated with thermal treatment to about 260 °C and rapid cooling, resulting in the production of an iron/tin alloy. Steel is electrolytically coated with a thin layer of metallic tin in a bath of tin sulfate in sulphuric during the tin coating process.

4.7. Environmental Impact of Metal-based Packaging

Metal packaging has both positive and negative effects on human health and the environment. The positive aspects of metal-based packaging are having already been discussed such as rigid structure, withstand high temperature, excellent gas, light and moisture barrier properties, etc. The negative aspects include global warming caused by carbon dioxide emissions from metal manufacturing units, the leaching of harmful toxic chemicals from containers into food, and resource depletion. Deshwal & Panjagari (2020) reported metal packaging waste accounts 10% of the total food packaging waste that is comparatively less than other packaging materials such as paper and paperboard, plastics, glass and woods. Better research and policies are needed to maintain a balance between both aspects of metal food packaging. Packaging waste recyclability is major concern in food processing industry. The consumption of packaged and processed foods had increased the proportion of packaging waste in total municipal solid waste mass. Steel's magnetic properties make segregation and recycling simple, and it has a higher recycling rate (around 60%) when compared to other metal packaging materials. For reducing metal packaging material waste and its negative effects, various management approaches/ pathways such as recycling, landfilling, and combustion for energy recovery are used. A tin can decomposes in 6–12 months after disposal and subsequent incineration, but it takes 6–10 years for complete reduction by oxidation without incineration (Griese 1971). Some Indian standards specification related to safety and recyclability aspect of metal packaging materials are given in Table 4.

Table 4: Indian Standards applicable in metal packaging

Indian Standard regulations	Year	Subjects
IS 2471	1963	Methods of test for metal containers
IS 6941	1973	Method for sampling metal closures
IS 2134	1981	Round tins for general purposes
IS 9991	1981	Condensed milk cans
IS 11079	1984	Metal cans for tobacco-based products
IS 1994	1987	Crown closures
IS 9396-1	1987	Round open top sanitary cans for foods and drinks
IS 5818	1988	Specifications for lacquers and decorative finishes for food cans
IS 3603	1988	Seamless aluminum bottles
IS 2087	1988	Square tins for general purpose
IS 2552	1989	Specifications of steel drums

Indian Standard regulations	**Year**	**Subjects**
IS 8970	**1991**	**Specifications of aluminum foil laminate for packaging**
IS 1406	1995	Rectangular tins for liquids
IS 3101	1995	Aluminum collapsible tubes
IS 14407	1996	Aluminum cans for beverages
IS 10325	2000	Square tins-15 kg/L for ghee, vanaspati, edible oils and bakery shortenings
IS 1993	2006	Cold-reduced electrolytic tinplate
IS 2034	2012	Round open top sanitary cans for butter and cheese
IS 11078	2012	Round open top sanitary cans for milk powder

Table 5: Some standard can designations, size and capacity

Can Designation	Size (Diameter mm X Height mm)	Capacity (ml)
No.1 Tall	77.8 x 119.1	426
No.1 Flat	87.3 x 56.6	216
No. 2 ½	103.2 x 119.1	850
No. 5	130.2 x 142.9	2330
No. 10	157.2 x 177.8	3110

Acknowledgement

The authors are thankful to the members (faculty/research scholars) of Department of Food Engineering, National Institute of Food Technology and Entrepreneurship and Management (NIFTEM), Kundli, Haryana, India for their supports during writing the book chapter.

References

Board, N. I. I. R. (2010). Handbook on modern packaging industries. Asia Pacific Business Press Inc.. PP 15-21.

Coles, R., McDowell, D., & Kirwan, M. J. (Eds.). (2003). Food packaging technology (Vol. 5). CRC Press.

Cvetkovski, S. (2012). Stainless steel in contact with food and bevarage. Metallurgical and Materials Engineering, 18(4), 283-294.

Del Pianta, D., Frayret, J., Gleyzes, C., Cugnet, C., Dupin, J. C., & Le Hecho, I. (2018). Determination of the chromium (III) reduction mechanism during chromium electroplating. Electrochimica Acta, 284, 234-241.

Deshwal, G. K., & Panjagari, N. R. (2020). Review on metal packaging: materials, forms, food applications, safety and recyclability. Journal of Food Science and Technology, 57(7), 2377–2392. https://doi.org/10.1007/S13197-019-04172-Z/FIGURES/5

Dey, S., & Agrawal, M. K. (2016). Tinplate as a sustainable packaging material: Recent innovation and developments to remain environment friendly and cost effective. Int. J. Res. IT Manag. Eng, 8, 9-22.

Global Metal Cans Market | 2022 - 27 | Industry Share, Size, Growth - Mordor Intelligence. (n.d.). Retrieved February 1, 2022, from https://www.mordorintelligence.com/industry-reports/global-metal-cans-market-industry

Griese EW (1971) Metals and plastics in packaging and the environment. J Milk Food Technol 34(5):232–235.

J. Davis (1999), Corrosion of aluminum and aluminum alloys (2nd ed.), ASM International, USA, pp. 25-49.

Jafari, S. M., Amanjani, M., Ganjeh, M., Katouzian, I., & Sharifi, N. (2018). The influence of storage time and temperature on the corrosion and pressure changes within tomato paste cans with different filling rates. Journal of Food Engineering, 228, 32-37.

Leimbach, M., Tschaar, C., Schmidt, U., & Bund, A. (2018). Electrochemical characterization of chromium deposition from trivalent solutions for decorative applications by EQCM and near-surface pH measurements. Electrochimica Acta, 270, 104-109.

P.W., B. (1978). Who introduced the tin can? Nicolas Appert? Peter Durand? Bryan Donkin? [Food packaging]. Food Technology. https://doi.org/10.3/JQUERY-UI.JS

Peter KTO, Ulrich N (2007) International Life Sciences Institute Report, Packaging Materials. Printed by ILSI, Europe

Robertson, G. L. (2005). Food packaging: principles and practice. CRC press.

Sarkis, A. M., Robin, A., Souza, V. A., & Suzuki, P. A. (2011). Woodgrain defect on tinned steel Flandres foil. Materials characterization, 62(6), 621-625.

Taroco, H. A., Garcia, E. M., Duarte, L. F., Madeira, A. P. C., Souza, A. G., Taroco, C. G., & de Fátima Gorgulho, H. (2020). Factorial Design Study of Corrosion in Tin-Free Steel (TFS) Tinplate of Tomato Paste. Journal of Packaging Technology and Research, 4(2), 135-144.

Wijenberg, J. H. O. J., Steegh, M., Aarnts, M. P., Lammers, K. R., & Mol, J. M. C. (2015). Electrodeposition of mixed chromium metal-carbide-oxide coatings from a trivalent chromium-formate electrolyte without a buffering agent. Electrochimica Acta, 173, 819-826.

Williams, G., & McMurray, H. N. (2009). Underfilm/coating corrosion. In Shreir's Corrosion (pp. 988-2004). Elsevier Science. 4th ed., 2, p. 988.

Wint, N., Warren, D. J., DeVooys, A.C.A., & McMurray, H. N. (2020). The use of chromium and chromium (III) oxide PVD coatings to resist the corrosion driven coating delamination of organically coated packaging steel. Journal of The Electrochemical Society, 167(14), 141506.

5

Glass Packaging

Twinkle Kumar Sachchan, Saumya Chaturvedi
Abhilasha Sharma, Anjali Bangar, Mansi Titoria

Shaheed Rajguru College of Applied Sciences for Women, University of Delhi Vasundhra Enclave, Delhi

5.1. Introduction

Glass is one of the oldest packaging materials used for food. The use of glass as a packaging material dates back to 3000 BC where it was the first object to be used for containing and carrying food objects. It is made by cooling a heated fused mixture of silicates, lime and soda. It is an amorphous material in which all the constituents individually form an irregular network at microscopic level. It means it has failed to form a definite crystalline structure during cooling. Glass is defined as "amorphous, inorganic product of fusion that has been cooled to a rigid condition without crystallizing (ASTM, 2010)". It has no definite melting point. It is chemically inert and can contain food for a long period of time. It is suitable for hot filling, thermal processing as well as chilled storage. Glass is best known for its thermal shock and hydrolytic resistance. Recycled glass is also used as packaging material and is more cost effective than virgin glass as it requires less energy for production. Utilization of glass in industries for food packaging has been significantly risen in recent years not only because of its inert nature and easy reusability but also to decrease the excessive use of plastic container as food packaging material (Bansal et al., 2013, Rasmussen, 2012).

History of Glass

Glass manufacturing has been traced back in 3200 BC in Mesopotamia. Beads were the earliest known glass objects during 2000 BC which were a product of accidental metal working. Initially glass ropes were coiled, shaped and fused together. From the time of discovery, the manufacturing process of glass and its components have changed only slightly but there has been substantial change in the molding techniques.

Glass jar industry originated in the early 1600s. The invention of automatic glass bottle blowing machine in 1880 led to industrialization in the process of bottle making. In 1903 Michael J. Owens at the Toledo, Ohio, plant of Edward D. Libbey designed and built first fully automated machine for making bottles. Then only glass containers of all shapes and size became economically viable for consumer products, and controlled the market from the initial 1900s till the late 1960s. The development of plastic packaging after world war II replaced a large share of glass packaging especially in beverage industry. However with rising global concern due to single use plastic, food packaging industry are reverting back to sustainable packaging material. Since glass is made from natural sources unlike plastic, which is made from synthetic material, glass is bound to dominate the global market for food packaging

Glass as a Food Packaging Material

Glass is very suitable packaging material for food products. It is impermeable to gases and moisture thus making it a suitable choice as a packaging material for food. Glass with non-toxic and non-porous surface has a high degree of impenetrability and also prevent leaching of toxins into the food. Glass's impervious nature makes them very tough to tamper with and hence provides tamper resistance. It retains delicate flavours of the food and also defends against spoilage reactions like oxidation. It is a brilliant packaging material for liquids that requires long storage, such as liquor or wine, as it can holds the flavour for long time. The suitability of glass jars to temperature variations and heating processes is also advantageous as compared to other packaging materials.

Ingredients of Glass

For glassmaking, the basic materials come from mines which are smelted. It means they are chemically reduced to their oxides at temperatures more than 1500°C. Silica is the key element for glass making and is obtained from sand, quartz or flint. To make fused silica glass, silica is heated at around 1723°C. Silica is mixed with other raw materials in different ratios in most of the glasses. Alkali fluxes (i.e., sodium and potassium carbonates) are added to reduce the fusion temperature and viscosity of silica. Magnesium and calcium carbonates (dolomite and limestone) are added to prevent the glass from dissolving in water and thus they act as stabilizers. Some ingredients are added to improve physical properties of the glass. For example, lead is added to provide clarity and brilliance, but it also softens the glass. Alumina is added for hardness and durability. To make borosilicate glass, 6% boron is added. Boron reduces leaching of sodium from glass. Sodium is loosely combined in silica matrix and hence has tendency to dissolve out from glass. Addition of boron thus prevent glass from corrosion.

Composition for soda-lime glass is mentioned in Table 1. The percentage of different components vary depending on end use of the glass and which raw materials are used, such as, silica (SiO_2) 68%-73%, soda (Na_2O) 12%-15%, calcia (CaO) 10%-13%, iron oxides (FeO) 0.05%-0.25% and alumina (Al_2O_3) 1.5%-2%. Glass prepared from soda-lime accounts for approximately 90% of all glass formed and are cast-off for the manufacturing containers which do not require extraordinary chemical strength and heat resistance (Gliozzo, 2017). Borosilicate glass is used as glass oven ware which is prepared by changing the alkali with boric oxide.

Table 1: Composition of Glass (Robertson, 2012)

Material	**Weight (kg)**
Soda ash Na_2CO_3	100
Sand SiO_2	300
Cullet	460
Aragonite $CaCO_3$	90
Feldspar (SiO_2.Al_2O_3)	40
Salt Cake NaCl	4
Total	994
Yield of glass	909

5.2. Classification of Glass

Based on their degree of chemical and hydrolytic resistance glass containers are classified into Type I glass, Type II glass, Type III glass and Type IV glass. Although such distinct classification is more necessary in pharmaceutical industry, food industry also takes a cue from such classification for various application.

5.2.1. Type I (Highly resistant, Neutral or Borosilicate glass)

It encompasses 80% silica, 10% boric oxide and sodium & aluminum oxide in small quantities. This type of glass is chemically inert and has high hydrolytic resistance because of boric oxide. Because of its low expansion coefficient, it has high thermal shock properties. This is suitable as packaging material for most preparations whether parenteral or non-parenteral. They can be utilized as containers for strong acids and alkali (Sayre, E. V. 2016).

5.2.2. Type II (Treated soda-lime glass or Soda-lime-silica glass or Dealkalized soda lime glass)

Type I glass are modified by treating its inner surface with sulphur to increase its hydrolytic resistivity. This results in the production of Type II glass. This helps in removing leachable oxides and leads to prevention of blooming and weathering from bottles. It is easily moldable as it has less melting point than Type I. Soda lime silica glasses are appropriate for most acidic and neutral aqueous preparations whether parenteral or non-parenteral (Berenjian, A., 2017; Gliozzo, 2017).

5.2.3. Type III (Regular soda lime glass)

Type III has average chemical resistance as it is an untreated soda lime glass. It has 75% silica, 15% sodium oxide, 10% calcium oxide, traces of aluminum oxide, magnesium oxide, and potassium oxide. Aluminum oxide is responsible for the chemical durability. Magnesium oxide reduces the molding temperature. They are used as packaging material for parenteral products or powders. They are used in packaging non-aqueous preparations and powders for parenteral use with the exception of freeze-dried preparations. It is also used in packaging non-parenteral preparations (Bidegarary et al., 2018).

5.2.4. Type IV (General-purpose soda lime glass or Type NP glass)

Type IV glass are not suitable for products that needs to be autoclaved because of their low hydrolytic resistivity and it also increases erosion reaction rate of the glass. It is used to store topical products and dosage forms (Vinci, 2019).

5.3. Manufacturing of Glass

The process of glass making can be divided into 5 stages as follow:

5.3.1. Collection of Raw Materials

Raw materials are selected on the basis of type of glass to be manufactured. Materials required for the production of different glass containers are as mentioned in the Table-2.

Table 2: Raw materials for each type of glass

S.no.	Glass Types	Raw Materials
1.	Soda-lime glass	Clean sand, Soda ash and Chalk
2.	Potash-lime glass	Clean sand, Potassium carbonate and Chalk
3.	Potash-lead glass	Pure sand, Potassium carbonate, Lead sesquioxide or Litharge
4.	Common glass	Salt cake, Chalk, Ordinary sand and Coke etc.

In addition to ingredients mentioned in Table 2 cullet and decolorizers are also added during manufacture of glass. Cullet helps in increasing the fusibility of glass and prevents alkali loss by vitalization during the reaction in forming new glass. Traces of iron compounds such as ferrous oxide and ferric oxide present in raw materials imparts green and very light-yellow tint to the glass. To avoid this colour formation decolorizer such as nickel oxide (NiO), manganese dioxide (MnO_2), cobalt oxide (CoO), arsenic oxide (As_2O_3) and antimony oxide (Sb_2O_3) are added (Berenjion, et al., 2017).

5.3.2. Batch Preparation

Raw materials, cullet and decolorizer after weighing in accurate proportions are finely powdered in the grinding machine. Powdered ingredients are put in mixing machine and a uniform mixture is obtained. This uniform mixture is called 'batch' or 'frit' and is taken for the melting procedure in a furnace.

5.3.3. Melting In Furnace

The prepared batch is melted in the furnace. Two types of furnaces are available i) Pot furnace ii) Tank furnace. The batch is heated till all the gases such as carbon dioxide, oxygen, sulphur dioxide escapes.

5.3.4. Fabrication

Fabrication is the process where a suitable shape is given to the molten glass. This can be done by three methods:

i) Blow and Blow

ii) Wide Mouth Press and Blow

iii) Narrow Mouth Press and Blow

(i) *Blow and Blow*

This method is used to make small mouthed bottles. Here two processes of blowing take place. The molten glass mass called a gob is formed into a parison by blowing compressed air. This results in neck formation giving gob a uniform structure. The parison is inverted and compressed air is blown through the mouth of the glass bottle to give its final shape (Figure1) (Dubé et al., 2010).

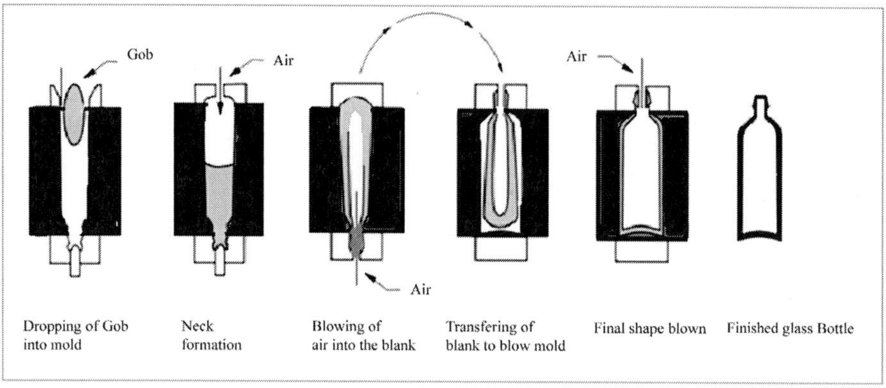

Fig. 1: "Blow and Blow" process for making glass bottles

(ii) Wide Mouth Press and Blow

This method is used in jars, where the first pressing process takes place, and after that blowing takes place. The gob is dropped into a mold having a plunger. The plunger compresses the gob into the desired shape parison of a wide mouth. It is then transferred to blow mold where its final shape is given by blowing compressed air into it (Figure 2) (Bansal et al., 2013).

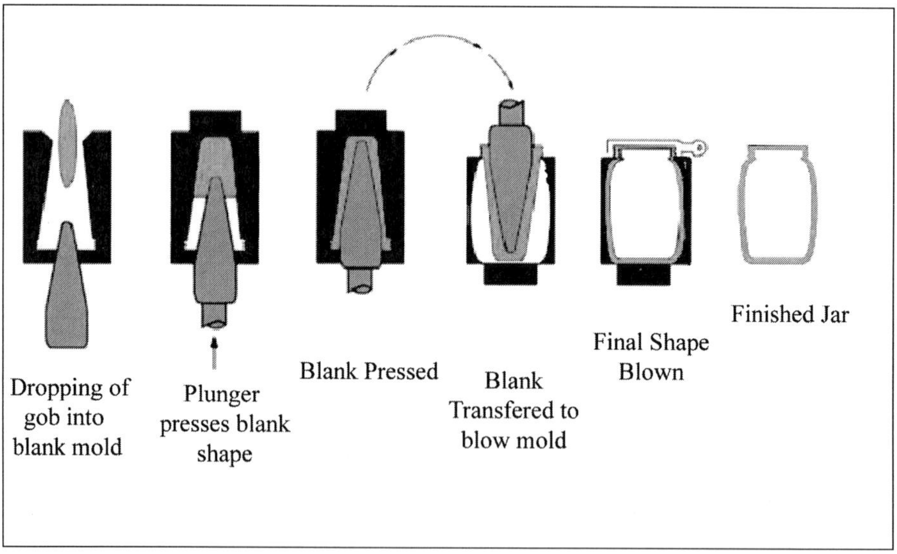

Fig. 2: Wide Mouth "Press and Blow" process for making glass jars .

(iii) Narrow Mouth Press and Blow

This is a more current procedure for making light weight bottles. Here pressing and blowing process take place. Light weight bottles manufactured by this method can withstand thermal shock of 5-10°C more than the regular glass bottles (Dubé et al., 2010). In this method of manufacturing narrow mouth glass bottles, a plunger is used for the neck formation of the parison instead of using compressed air.

5.3.5. Annealing

Annealing refers to removal of stress. During the glass formation process there is often the development of certain stresses in various regions of the glass mainly due to non-uniform cooling. Annealing is done to reduce the stresses in critical regions of glass containers thereby enhancing the stability. The glassy solid undergoes relaxation as it is cooled. Sufficient time should be given for relaxation in order to reduce internal stresses. Annealing is attained by holding the glass product at around 5 °C (9 °F) above its annealing point for 5 to 15 minutes, followed by slowly cooling it through the glass-transition range and the strain point and finally to room temperature. Annealing of glass is the process of slow and homogeneous cooling of glass articles. If glass articles are allowed to cool down rapidly, due to bad conductance of heat its superficial layer cools down first and interior part remains comparatively hot. Due to this glass remains in a state of strain. Therefore, this type of glass is easily breakable, even with a light shock or disturbances (Radchenko et al., 2020).

5.4. Strengthening of Glass

The strength of the glass product may be increased by using certain processes that can provisionally reduce the intensity of flaws.

5.4.1. Polishing and Glazing

A solution of 6-30% hydrofluoric acid with small amount of sulfuric acid (for safety purpose) is used for etching process in most silicate glasses. This treatment is not recommended because polishing leads to removal of the finest layer of glass present in the proximity of glass surface. This reduced glass thickness has deleterious effects on structural integrity of the glass. A thin layer of another glass having lower thermal expansion property than substrate, is fired onto the glass product. This strengthening process is called overlay glazing. In contrast to polishing, overlay glazing strengthens the glass by its increasing its thickness.

5.4.2. Thermal Tempering

Thermal tempering is accomplished by the process of quenching or known as rapid cooling where the temperature of the glass is higher than the transition range (glass transition range can be defined as a temperature range in which amorphous regions of glass undergo reversible transition forming a brittle and glassy state) by utilizing symmetrical air jets. The outer layer of the glass cools faster and thus have higher shrinkage rate than inner layer of glass. The outer layer is later contracted and inner layer is stretched. This increases the durability and scratched resistance so that glass doesn't loses its functionality even at adverse conditions.

5.4.3. Ion Exchange

Only alkali containing glasses are suitable for ion exchange strengthening. In this method, glass is immersed in molten alkali salt (mostly nitrate) at temperatures below transition ranges (75 to 150K).

5.5. Merits and Demerits of Glass as Packaging Material

Glass is very well attested for being the least toxic packaging material for foods and drinks. The precedence of glass as a packaging material is because of its imperviousness (maintaining the freshness of the foods for a longer duration without undermining the taste or duration), inertness, resistance to tampering, hygiene, design, quality colour, transparency, decoration potential, chemical propriety, heat treatability, and microwave ability (Girling, 1999). Glass also provides good insulation. It does not get corroded and is also available in various shapes and sizes. Fancy appearance of glass containers enhances the visual appeal of the product by providing it transparency. Glass is transparent or translucent because it has no mobile electrons to absorb light.

Glass can withstand high processing temperatures making it ideal for thermal sterilization processes of both neutral and acidic foods. Glass bottles are most predominantly used for soft drinks, alcoholic drinks, and portable water. Usability of colored glass bottles is high in oil and oil-based products so to retard rancidity caused by UV rays. Glass is also used for canning of delicate and superior quality food products so as to provide both preservation and visibility.

Fig. 3: Variety of Glasswares for food packaging

Due to the glass's impermeable structure and body, no chemical diffusion of the contents take place. This also implicates to the compounds which are on the outside i.e., printing inks, adhesives etc.

- ***Inert Nature*:** Glass is used as a packaging material because of its chemical structure which is an amorphous non-crystal solid with very small pore size which doesn't allow any atoms or molecules to pass through it, hence making it inert in nature.
- ***Barrier Properties*:** Glass bottles have barrier properties which prevent the invasion of contents by oxygen and other gases, and at the same time can prevent the volatile components of the contents from volatilize in the atmosphere.
- ***Thermal Strength*:** An ability to withstand any sudden temperature change by glass is known as its thermal stability. This property is of major noteworthiness in the food industry applications. Some glass containers are used in the application of high temperature processing.

Glass containers are found to exhibit various types of resistance during processing which are important for the production of carbonated beverage etc. where the glass container being used is expected to be processed in boiling water or pressurized hot water and it results in production of constraints at various points on the outer surface of the container which is known as the internal pressure resistance. Glass containers should also exhibit resistance to impact, and it is very essential in two aspects i.e., when a moving container is contacting a stationary object as when a bottle is dropped or a moving object contacting a stationary bottle (as in the filling line). The design features should be subsumed in such a way into the sidewall so as to strengthen the contact points. Surface treatments should also be incorporated to minimize the fragility of glass when it might contact a immobile object. The comprehensive toughness of the glass can be significantly compromised due to surface damage such as abrasions and scratches hence resistance to scratches and abrasions is pivotal.

A major advantage of using glass as a packaging material is to circumvent and prevent product shunning and bio terroristic attacks. Thus, giving glass its inviolability i.e., glass products cannot be perforated by syringes and making this an integral factor for major food companies (Le Bourhis, 2014).

However, usage of glass as a packaging material for various foods and beverages can prove to be a limitation or a drawback as a type of food packaging material. Regardless of the efforts of using thinner glass in the packaging, its heavy weight adds up into the transportation costs and therefore adding up in the costs for the consumers too (Vinci, 2019).

The glass containers fracture in temporary stresses which may arise due to sudden temperature fluctuation i.e. sudden heating or cooling and strong impact can cause fragility of the glass containers making it futile and misspending its contents. Food companies usually secure food inside a one-piece plastic pouch or a metal can and rest is left out to the consumer to get the food out of the container but with glass packaging food companies have to add an additional thing to seal the container shut for example cork, lid or a bottle cap and thus manufacturers and bottlers have to take integral steps to prevent any contamination from entering. Glass is also much more unsafe when broken as a broken piece of glass can cut into a person's skin and at times be life threatening and it possess a risk for those handling during the manufacture, in shops, during transportation and also those responsible for packaging. Many ions like sodium can also itinerate from glass to food under certain conditions (Gallucci et al., 2020).

High quantity of water is needed for the manufacture of glass which would give the water footprint of glass specifically. This could be a factor of curtailment in areas having low water availability (Wohner et al., 2020).

5.6. Testing of Glass Containers

Glass is best known for its thermal shock and hydrolytic resistance. Quality of glass containers are evaluated by certain tests which are set under the standards of appearance, consumer usability and compatibility and chemical testing. These tests include whole container test, crushed glass test, leakage test, thermal shock resistance test, chemical resistance test and water attack test.

The specifications of these tests are set according to the guidelines provided by regulatory bodies such as WHO, USFDA, ICH and GMP.

***Appearance Tests**:* This involves evaluating glass containers on the basis of critical, major or minor defects. Critical defects are extremely hazardous to the consumers, major defects reduce the usability of contents packed in glass container and minor defects do not possess any risk to consumer health but only affects the physical appearance of the containers.

Dimensions Tests: Measurements of dimension of a glass container are taken to ensure uniformity of the containers. Variety of measuring equipments such as callipers, screw gauges, micrometers etc are first calibrated and then used for precise and accurate results.

***Consumer Usability and Compatibility*:** Ensures adequate fitting of all the parts of container formed and their appropriate functioning. The cap or nozzle of the glass containers / bottles must be perfectly fitted and consumer friendly.

***Chemical Testing*:** Principle involved in different chemical tests are mentioned below:

Crushed Glass Test: The glass containers is crushed and then sifted to obtain definite size particles. Constant surface area of glass is ensured by controlling the particle size and weight of powder. Neutral or surface treated glasses are distinguished by the application of this test.

Water Attack Test: The basic principle of this test revolves around acid-base titration with the application of methyl red as an indicator. The test is specific for containers (Type II) which are being treated with sulphur dioxide fumes under controlled humidity to neutralize surface alkali. This test determines whether the alkali leached from the containers is within the specified range or not by treating it with acid under elevated temperatures.

Whole Container Test: The test is valid for soda-lime treated containers only. The containers filled with test solutions are exposed to specific environmental conditions. Small sized containers are more affected by alkali leaching from surface as surface area remains constant and volume increases with increase in container size (Khan et al., 2010).

Powdered Glass test: The test determines the amount of leached alkali from powdered glass by titrating it against 0.02 N sulphuric acid and methyl red indicator.

Thermal Shock Resistance test: In this the glass material is heated in a hot water bath and then is plunged into a cold-water bath. The process is repeated until failure or breakage occurs, each time the repetition of test is done at 10°C higher temperature gradients. The test is significant for sensitive materials that can survive up to 42° C thermal gradient only. Thermal shock resistance is determined as the highest temperature gradient a container can survive without breaking.

Leakage Test: The glass container is filled with mythelene blue coloured solution and then it is autoclaved under pressure for 10-15 minutes at 121 °C. The container is evaluated on the basis of the colour retention.

Internal Bursting Pressure Test: American glass research increment pressure tester is the instrument used for this test. In this test, the glass bottles are filled with water and are placed in test chamber. The application of scaling head results in increase of internal pressure gradually in series for a certain period of time until the bottle bursts.

5.7. Recycling of Glass Packaging Material

Recycled glass is used more than the virgin glass as it requires less energy for production and is more cost effective. Great recyclability of glass has indispensable benefits as it merely needs to be re-melted and reformed which in turns reduces energy demand to produce another glass article. As very little waste is generated during glass recycling and glass manufacture processes, the recycled glass product can be conveniently used again and again. In comparison with other packaging material glass is easily recyclable all and they can be recycled all together because it is made from a single material and there is no mixture of materials to separate. Before the recycling process is initiated, the glass material has to be sorted and separated as this is a vital step in the process of glass recycling which allows segregation of major and minor contaminants. These contaminants are not glass materials but have found their way among the glass. Glass packaging's carbon footprint can be dramatically lowered down through the use of recycled glass. Cullet i.e furnace-ready recycled glass material is free of contaminants such as ceramics, metals, stones, or gravel, etc. and 80% of the recovered glass containers can be recycled. Recycling of glass packaging material also has a great effect on the environment conservation and resources protection (Ran et al., 2016).

5.8. Defects in Glasswares

Glass, despite of being premium packaging material for food products could be unfit as well because of the defects observed during manufacturing of glass bottles. Development of sub-microscopic defaults during the cooling of glassware leads to weakening of the container. The outcome of these defects in glass container would be generation of glass shards which in turns contaminate the food product packed in that container. When a force in form of tensile stress is applied, flaws propagate resulting in the weakening of glass walls. These walls are susceptible to breakage and are a ramification of critical flaws of the glass containers. Such flaws can be present during the manufacturing, filling processes, transportation, or storage. Glass defects can also be characterized as: Defects caused by contamination, mechanical, or chemical influences in or on the surface (Miller et al., 2001).

Defects could be minor neither harming the product nor consumer or major possessing potential safety hazard. Minor defects include spike, neck spike, deformation, pits on the surface, chipped finish, stone in glass, sagging surface, bent neck etc. Whereas Major defects encompass internal and external blisters, internal fragments, stuck wares, burrs, bird swing, checked, cracked seams etc.

Minor Defects: *Spike* is defect in which bottom on the inside of glassware has a protrusion extending upwards while in *neck spike* defect is due to irregular wall thickening near neck region of the glass bottle. In *deformation* and *bent neck,* the article is malformed or tilted from one end because of which packaging of any product in such defected containers is unattainable. *Chipped finishing* involves minute glass fragments flaked off either from top or from sides of container. *Stone in glass* is an opaque unmolten matter imbedded between the glass walls. Surface defects such as *Pits on the surface* and *sagging surface* only affects the appearance and not the usability of the container. Formation of scales are observed in sagging surface (Sun et al., 2020).

Major Defects: *Blisters* can develop internally as well as externally. This is due to entrapment of air bubbles within the molten glass. They may be cracked or uncracked. Blister posses' high risk of damage to the operators as well as consumers during handling. *Internal glass fragments* whether attached or loose inside the glass jar contaminates the food and is a possible hazard to consumer safety. *Burrs* includes development of sharp edges along the sides of mould seams and posses' risk of injury to the manufactures. Likewise, when two glassware comes in contact during manufacturing, appointed cutting edge is formed at the site of their contact known as *Stuck ware. Bird Swing* is a thin glass splinter inside the glass which is difficult to detect in filled containers. A feather like crack on the glass article is called *Checked. Cracked seam* is a hairline fracture in glassware (Miller et al., 2001).

Major faults not always lead to breakage of the article but weakens the walls and sometime causes generation of sharp pointed edges or fragments possessing great physical hazard to both the operators and consumers. In contrast, minor defects affect the visual characteristics of the article and is not a concern to public health (Willhoft, 1986).

5.9. Current Status of Glass Packaging Industry

Utilization of glass in industries for food packaging has been significantly risen in recent years not only because of its inert nature and easy reusability but also to decrease the excessive use of plastic container as food packaging material. Glass properties which have driven its market to forecast period are long-term preservation, reusability, inertness and innovative designs.

As per Androit market research 2019, fast moving consumer goods sector is the major consumer of glass packaging industry (Figure 4).

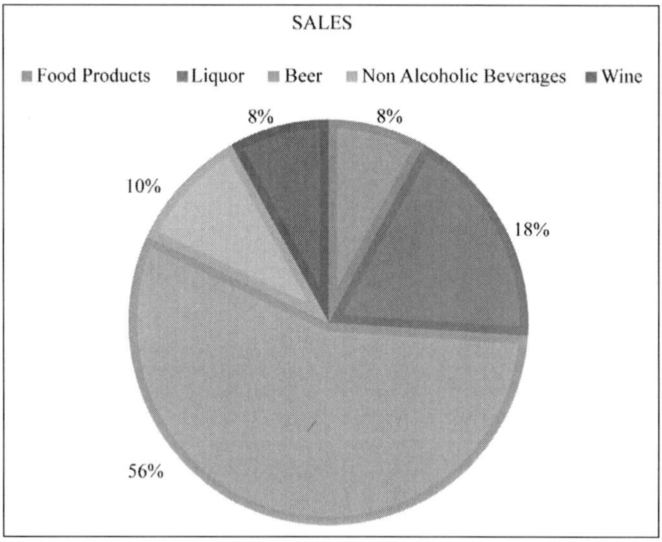

Fig. 4: Global sales volume of glass containers, 2018

China and India are major buyers of FMCG products. Globally the glass packaging market is anticipated to grow with CAGR (compound annual growth rate) 4.39% over the forecast period of 2020 – 2027. Asia pacific has largest market for glass packaging geographically whereas Europe has fastest growing market.

As per the recent analysis report by Mordor intelligence 2021, India is about to witness a substantial growth in glass packaging market of about 6.94% CAGR during the forecast period of 2020-2025. Beverage sector is anticipated to drive the growth of market sector because according to CARE ratings the

Indian alcohol market is expected to witness significant growth about 18.91% till 2021.

5.10. Innovations in Glass Packaging

There is an evolution observed within the glass industry due to the ban of plastic around the globe. The demand of glass as a packaging material increased many folds. With this increasing demand for glass, the outlook and the perception of the glass packaging industry has changed and impacted the major developments in the food industry (Romani, 2002).

5.10.1. New Lightweight Glass Packaging

Keeping in mind the long-term aspect of the packaging material, lightweight glass making technique was introduced which enabled the customer greatly in terms of carrying the product safety without any danger of breakage. It has also helped with the concept of minimal storage.

5.10.2. Customized Glass Bottles and Containers

Various machines are introduced in the industry which are highly capable of manufacturing a wide range of glass bottles with various shapes and designs.

5.10.3. Glass Cutting Edge of Technology

Modern techniques have helped in imparting various necessary properties to the glass material such as preventing it from packaging corrosion and providing thermal shock resistance (Gallucci et al., 2020). Also, it helps in transmission and absorption of light making it a type of smart glass and giving it the anti-reflection properties.

5.10.4. Environment- Friendly Packaging/ Ecology Bottles

Tons of natural resources are saved with the recycling of glass which has reduced the carbon dioxide emission by a ton. This has given an environment-friendly approach to glass manufacturing processes (Vinci, 2019).

5.10.5. "R" Mark Bottles (REUSE)

These bottles are specifically designed for the reusable purpose of packaging material and is very significant in the conservation of natural resources as well as the environment. They are designed in a specific way that ensures their reuse after the actual product consumption (Vinci, 2019).

References

ASTM. 2010. Standard Terminology of Glass and Glass Products, C162–05. American Society for Testing and Materials. For the latest version of each standard visit www.techstreet.com.

Bansal, N. P. & Doremus, R. H. (2013). Handbook of Glass Properties. Elsevier.

Berenjian, A. & Whittleston, G. S. (2017). History and manufacturing of glass. American Journal of Materials Science, 7(1), 18-24.

Bidegaray, A. I., Ceglia, A., Cicconi, M. R., Pham, V. T., Crabbé, A., Nys, K., Godet, S. (2018). An in-situ XANES investigation of the interactions between iron, manganese and antimony in silicate melts. Journal of Non-Crystalline Solids, 502,227-235.

Dubé, M., Doquet, V., Constantinescu, A., George, D., Rémond, Y., &Ahzi, S. (2010). Modeling of thermal shock-induced damage in a borosilicate glass. Mechanics of Materials, 42(9), 863-872.

Emblem,A.(Ed.).(2012). Packaging technology:Fundamentals,materials and processes. Elsevier.

Farmer, N. (2013). The future: global trends and analysis for the international packaging market in relation to the speed of impact of packaging innovation and likely material changes. Trends in Packaging of Food, Beverages and Other Fast-Moving Consumer Goods (FMCG), 288-312.

Gallucci, T., Lagioia, G., Piccinno, P., Lacalamita, A., Pontrandolfo, A., & Paiano, A. (2020). Environmental performance scenarios in the production of hollow glass containers for food packaging: an LCA approach. The International Journal of Life Cycle Assessment, 1-14.

Girling P.J. 1999. Packaging in glass bottles. In: Handbook of Beverage Packaging,Giles G.A. (Ed.). Boca Raton, FL: CRC Press, pp. 53–70.

Gliozzo, E. (2017). The composition of colourless glass: a review. Archaeological and Anthropological Sciences, 9(4),455-483.

Le Bourhis, E. (2014). Glass: mechanics and technology. John Wiley & Sons.

Moretti, C., & Hreglich, S. (2013). Raw materials, recipes and procedures used for glass making. Modern methods for analysing archaeological and historical glass, 1,23-47.

Radchenko, P., Batuev, S., & Radchenko, A. (2020, December). Investigation of the glass barriers fracture under shock loads using the JH-2 model. In AIP Conference Proceedings (Vol. 2310, No. 1, p. 020267). AIP Publishing LLC.

Ran W, Chen F, Qianni Wu, Sen Liu (2016) A Study of the Closed-Loop Supply Chain Coordination on Waste Glass Bottles Recycling. Modelling and Simulation in Operations and Complex supply chains Volume 2016

Rasmussen,S.C.(2012).How glass changed the world:The history and chemistry of glass from antiquity to the 13th century. Springer Science & Business Media.

Robertson, G. L. (2012). Introduction to food packaging. Food Packaging: Principles and Practice, 3rd ed.; CRC Press, Taylor & Francis Group: Boca Raton, FL, USA, 1-8.

Romani, A., Scaringi, L., & Sciullo, A. (2002). Glass as packaging material. Industriedelle Bevande (Italy).

Sayre, E. V., & Smith, R. W. (2016). Some materials of glass manufacturing in antiquity. In Archaeological chemistry (pp. 279-312). University of Pennsylvania Press.

Schaut, R.A., Peanasky, J.S., DeMartino, S.E., Schiefelbein,S.L. (2014). A new glass option for parenteral packaging. PDA Journal of Pharmaceutical Science and Technology, 68(5), 527-534.

Sun, X., Yang, J., Qiu, P., & Hu, Y. (2020) Glass fissure and glass bubble defect detection in industrial pipeline in quality control.

Vinci, G., D'Ascenzo, F., Esposito, A., & Musarra, M. (2019). Glass Beverages Packaging: Innovation by Sustainable Production. Trends in Beverage Packaging, 105-133.

Wohner, B., Gabriel, V. H., Krenn, B., Krauter, V., & Tacker, M. (2020). Environmental and economic assessment of food-packaging systems with a focus on food waste. Case study on tomato ketchup. Science of the Total Environment, 738,139846.

6

Shelf Life of Packaged Food

Harpreet Kaur Jambh

Post- harvest Machinery & Equipment Testing Centre, ICAR-CIPHET, Ludhiana Punjab, India

6.1. Introduction

Food packaging has a significant contribution in preserving the quality of the food throughout the processing and storage. The package containing food should maintain the quality of the product for the longest time under particular environmental conditions. The quality of various food and beverages decline with the passage of time and food become unacceptable to consume. The time in which food deteriorates and become unacceptable is termed as shelf-life. The degradation of quality with time under storage conditions depends upon intrinsic as well as extrinsic factors. The shelf-life study of a product provides important information to manufacturer and consumer regarding the quality of the food product throughout the storage time.

6.2. Shelf Life

The quality of food products reduces with passage of time and a times comes when a product is not accepted for consumption. The time from manufacturing a product to its unacceptability is known as shelf life. The shelf life of a particular product is decided by the processer or manufacturer. Manufacturers usually provide the longest possible shelf life to the food material as ineffective shelf life will ultimately cause dissatisfaction to consumer which affects the sale and brand name of the product. Therefore, manufacturers are always cautious about the shelf life of the food.

The Institute of Food Science and Technology (IFST) Guidelines (1993) has defined shelf life as: "the period during which the food product will remain safe; be certain to retain desired sensory, chemical, physical, microbiological and functional characteristics; and comply with any label declaration of nutritional data when stored under the recommended conditions." In simple words, the shelf life is the duration in which quality parameters of the food product are adequate for consumption. Therefore, shelf life indicates the period for which food material can stay on the retailer's shelf followed by consumer's shelf till its quality deteriorates and becomes unacceptable.

In the EU, shelf life is not stated in law and there is not any legislation about evaluation of shelf life. As per the Directive 92/59/EEC on general product safety, the manufacturer is responsible only for putting safe products in the market. The EU Directive on food labelling (79/112/EEC) requires prepackaged foods to bear a date of minimum durability (best before or best before the end) or, for highly perishable foods from a microbiological point of view, use-by date. The date of minimum durability is described as 'the date till the foodstuff retains its specific characteristics under suitable storage conditions'. This date (shelf life) is specified under the responsibility of producer. The decision is based on requirement of food use by or best before date that falls under the authority which is liable for labelling as it is in the best place to evaluate the shelf life of particular food. It is considered an offence to sell any consumable article after the indicated use-by date.

6.2.1. Critical Descriptors and Indices of Failure

During the storage of food, significant changes occur that can be described by descriptors. The descriptor that regulate the shelf life of a food is termed as a critical descriptor. In designing of packaging material for food, it is necessary to describe the quality characteristics that define the acceptability of food to the consumer and is known as critical descriptor or indices of failure (IoFs). There are several IoFs for testing the shelf life of the product that determines the sample failure. The buildup of rancid flavor in cereals and deterioration of red color (bloom) in chilled beef due to oxidation are examples of IoF.

The first step is to define the IoFs of a particular food. Further step is to calculate the degree of the degradation by that IoF. The final step is to determine which IoF is affected by the packaging material as packaging cannot stop or slow the deteriorative reactions in food. For instance, how much amount of oxygen reacts with food causes oxidation that leads to the unacceptability of the product. The packaging material with different O_2 transmission rate also influences the oxidation reaction in food.

6.2.2. Cutoff Point

After identification of critical descriptor or IoF, the further phase is to determine the cutoff point (COP) or the endpoint of specific degradation, for instance how much O_2 or moisture reacts with the product to become unacceptable.

The COP indicates the limit beyond which the acceptance of the product by the consumer is declined significantly. A COP could be a microbial degradation of the product that makes it unsuitable for human consumption. The change in odor, texture, color, flavor, etc. that affects the acceptability of the product to the consumer is also considered as COP. Therefore, COP is a condition in which any chemical physical, microbiological or sensory parameters become unsatisfactory to the consumer and the time in which the food material shows such conditions is known as the shelf life of the respective material.

The sensory COP is determined by the descriptive and affective method. In a descriptive method, the quantitative and/or qualitative attributes of the product are measured by the trained panelist while the affective method does not require a trained panelist. For sensory analysis, a consumer panel would consider a suitable method to evaluate the quality and shelf life of a food material but frequently gathering consumer panels for several measurements would not be feasible and is expensive. However a sensory panel is more suitable for continual evaluation, its result would be more logical and not necessarily representative of consumer feedback. The correlation of data obtained from a consumer panel with that of trained panel assess the quality and shelf life of a food product which assist in sensory quality control programs.

6.3. Factors Affecting the Shelflife of Packaged Food

The shelf life of food is affected by following factors:

- Intrinsic factors: Product attributes resulting from the formulation and processing factors
- Extrinsic factors: The factors resulting from the environment in which the food product is exposed during the life
- Characteristics of the package

6.3.1. Intrinsic Factors

6.3.1.1. Water Activity

Water activity (a_w) is the ratio of the water vapor pressure of food material to the vapor pressure of pure water at a given temperature.

$a_w = p/po$ where p is the vapor pressure of water exerted by the food material and p_o is the saturated vapor pressure of pure water at the given temperature. Most of the fresh food has a_w above 0.99.

The microorganisms will not form spore, grow or generate toxic substances below its limiting value of a_w. The amount of water present can affect the chemical reactions in various means. It could behave as a solvent or reactant (as in sucrose hydrolysis), where it could exert a dilution effect on the substrates, decreasing the rate of reaction. Water also affects the viscosity of the food by forming complexes or hydrogen bonds with the reacting species. Thus, a_w has a crucial role in regulating undesirable chemical and enzymatic reactions. The rate of change in the reactions can be accelerated or minimized over a wide range of a_w as shown in Figure 1. A slight change in the value of a_w causes a substantial change in reaction rates.

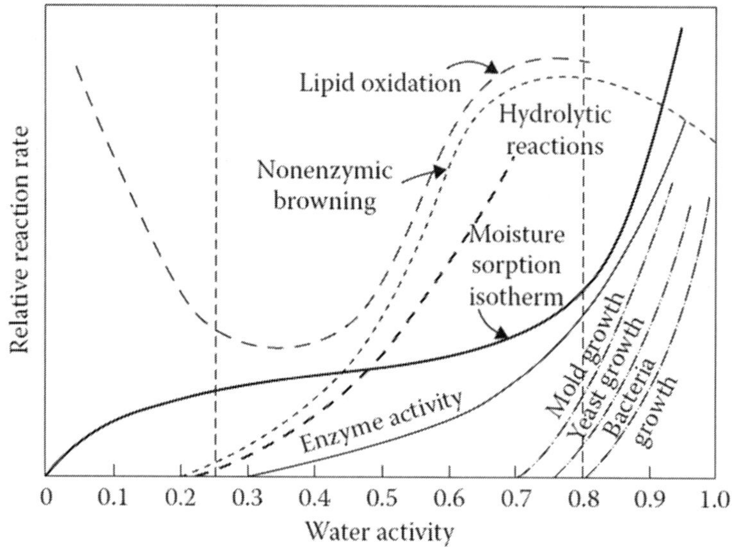

Fig. 1: Reaction rate as a function of water activity (Rockland & Beuchat, 1987)

6.3.1.2. Redox Potential

The oxidation-reduction potential (also called redox potential) (E_h or ORP) of a substrate is described as the ease with which the substrate gain or lose electrons. The more highly oxidized a substance, the more positive is electrical potential. It evaluates the oxidizing or reducing attributes of the food and depends on the composition of the food, temperature, pH, and amount of dissolved oxygen (DO).

In food, the reducing conditions are maintained by sulfhydryl groups on proteins, ascorbic acid and reducing sugars present in fruits and vegetables. E_h value can be changed by deteriorative reactions that takes place during the storage of the product. E_h plays a vital role in the growth, enzyme expression and thermal resistance of microorganisms. Aerobic microorganisms require positive values of E_h (oxidized) for growth whereas anaerobes need negative value of E_h. Various oxidation-reduction reactions takes place in milk during processing and storage. E_h and DO levels in enriched milk causes the oxidation of unsaturated fatty acids and the reduction in viability of probiotic strains such as bifido bacteria. Reduction in E_h could improve the quality of milk. The electro reduction of milk by membrane electrolysis shows that the electrochemical process reduced the E_h of milk without altering the nutritive and organoleptic values (Schreyer et al., 2008).

6.3.1.3. Availability of Oxygen

Oxygen is present within the environment of the food material and reacts with the components of the food. This leads to the deterioration of color, flavor, nutritional content, etc. of the food. Thus, the oxygen limits the shelf life of the food. But in some cases, the presence of oxygen is necessary to bring the desired characteristics in the food.

Food comprising high-fat content become rancid in the presence of oxygen. The oxidative rancidity causes a change in flavor of the product. The antioxidants, either naturally present or are added, slow the rate of reaction and delay the onset of rancidity. The reduction in the amount of oxygen present in the headspace or dissolved form and incorporation of antioxidants increases the shelf life of the food. Snack's products, for example, nuts are prone to rancidity and are packed with 100% nitrogen to preserve from oxidation and acts as a barrier to protect from physical damage. The packaging material such as PVdC-coated OPP/LDPE laminates are commonly used for food with short shelf life, whereas the food with higher value and longer shelf life are packed in metallized polyester/LDPE laminates.

6.3.1.4. Enzyme Activity

Enzymes act as biological catalysts that accelerate the reaction rate. The activity of enzymes within the packaged food is influenced by the existing situations within and outside the food. Controlling the conditions are essential to regulate the activity of enzymes. The factors such as water activity, temperature, pH, chemical preservatives, alteration of a substrate, composition of food and processing conditions control the activity of enzymes during storage. Among these, temperature, a_w and alteration of a substrate are the three most important factors in the context of packaging. Packaging material can maintain a low temperature of the product which results in retardation of enzymatic action and enhances the shelf life of the food. The low water activity means less availability of water that restrict the enzyme reactions. The alteration of the substrate such as the ingression of O_2 into a package is vital in the oxygen-dependent reaction that is catalyzed by the enzymes.

6.3.1.5. Microbiological Count

Under suitable conditions, the microorganism can grow and multiply that causes changes in the food that may or may not be desirable. Like in the case of fermentation, these are desirable otherwise it causes spoilage of food thus, food becomes unfit for human consumption. The microorganism that grows in the food require various nutrients such as water, energy source, nitrogen, vitamins, minerals, etc. and produce metabolic by-products in the form of gases or acids. The availability of water is directly related to a_w of the product. The primary source of nitrogen for microorganisms is amino acids. Extracellular enzymes such as amylases, lipases, proteases are also released by the microorganisms that affect the color, texture, flavor of the product. Some of the enzymes are present even after the death of the microorganism causing spoilage of the food. The thermal treatment is given to kill the microorganism that increases the shelf life of the product. Low acids foods are filled in hermetically sealed containers sterilized at 115.5 to 121°C or above to destroy all pathogens particularly *Clostridium botulinum*. The desired shelf life of the food is attained by both packaging material and pack seals that must provide a protection against microorganisms.

6.3.2. Extrinsic Factors

Extrinsic factors include temperature, RH, light, composition of gas in the packaging material influences the rate of deteriorative reactions in the product during storage and handling.

6.3.2.1. Temperature

The rate of deteriorative reaction is highly influenced by the temperature. The packaging material affects the temperature of the product. The deterioration reaction is caused by microorganisms that grow and multiply over a varied temperature range. Based on the temperature requirement for growth, microorganisms are divided into three groups.

(i) **Psychrotrophs**: The microorganisms grow well at or below 7°C with optimum temperature of 20 to 45°C. *Lactobacillus, Micrococcus, Pseudomonas, Psychrobacter, Enterococcus* are few examples of psychrotrophic species. These organisms also grow at refrigeration temperature results in decay of fish, meat, eggs, poultry, etc. stored at this temperature.

(ii) **Mesophiles**: These grow well at a temperature between 20 and 45°C with optimum temperature of 30°C and 40°C. *Enterococcus raecalis* is mesophile bacteria.

(iii) **Thermophiles:** These microorganisms grow well at and above 45°C with optimum temperature between 55°C and 65°C. Most important thermophilic bacteria in foods are belong to the genera *Bacillus* and *Clostridium*.

Like bacteria, molds are also able to grow over wide range of pH, osmotic pressure and nutrient content. Many molds, for example, *Aspergillus*, *Cladosporium*, and *Thamnidium* can grow at refrigerator temperatures. Yeasts grow over the temperature ranges of psychrotrophic and mesophilic but not within the range of thermophiles. The success of storage temperature depends upon other factors, for instance, RH and the absence or presence of gases. The packaging material must have insulating properties for the packaging and storage of chilled and frozen food.

6.3.2.2. Relative Humidity

The RH of the storage environment is a significant factor that influences the a_w and growth of microorganism unless a packaging material provides an excellent moisture barrier. Food products that suffer surface spoilage due to molds, yeasts, or certain bacteria should be stored in low RH conditions. In such casc, flcxiblc packaging material should be used that provides a barrier to moisture.

6.3.2.3. Gas Atmosphere

The storage of food in the environment containing a different proportion of gases impacts the growth of microorganisms. The composition of gases in the package is altered in a way to improve the shelf life of the food. Many deteriorative reactions occur in presence of atmospheric oxygen that affects the nutritional quality of the food. Hence, it is necessary to prevent the supply of O_2 inside the package. The removal of O_2 in vacuum packaging creates unfavorable conditions for the growth and development of aerobic microorganisms. The increased concentration of CO_2 in modified atmosphere packaging (MAP) retards microbial growth and extends the shelf life. The sealed glass and metal containers also inhibit the exchange of gases between food material and the atmosphere.

6.3.2.4. Light

The exposure of food product to light accelerates the deteriorative changes such as discoloration and change in flavor. Light, in visible range, lies between the wavelength of 4000 and 7000Å; whereas ultraviolet (UV) light lies between 2000 and 4000 Å. The catalytic effect of light is more in UV range and lower wavelength of the visible spectrum. The catalytic effect of light on the free radical reactions involved in fat oxidation not only lowers the nutritional value of the fat but also produces toxic compounds from the fats and oils and destroy fat-soluble vitamins, in particular vitamins A and E. The packaging materials can be modified by the application of a coating that absorbs light of a specific wavelength. The plastic films are incorporated with nano-sized particles of titanium dioxide that absorb UV-A and UV-B rays. Glass is altered by the addition of color-producing agents or by applying coatings.

6.4. Properties of the package

The package is selected based on the degree of protection it provides to the food from many extrinsic factors and affects the shelf life of the product. The material should be selected based on properties that would provide the necessary barrier for gases, light, moisture, etc. The major gas in packed food is O_2 that has an important role in various reactions such as microbial, oxidation, color change, etc. and results in reduction of the shelf life of the product. The packaging can control the amount of O_2 present in the product. The metal and glass containers are impermeable to odors, gases and water vapors whereas paper-based packaging material is considered as permeable. The protection provided by plastic-based packaging material depends upon the type of polymer used in its manufacturing. If a limited amount of O_2 is present in hermetically sealed metals and glass containers, the reaction cannot exceed a level that causes any detrimental changes in the product. The light damages the nutritional content of the food affecting the light-sensitive vitamins (vitamin A, B_2, D), color and taste. In this case, packaging material has properties that provide a barrier to the light.

The packaging material that protects against gases, water vapor, light, etc. is in direct contact with the food may serve as potential agent for physicochemical migration of ions and molecular species from packaging material to the food that might be harmful to the human and also restrict the shelf-life of the food. The residual monomers and plastic additives such as solvents and plasticizers present in the polymer are in close contact with the food. Additives are used to alter the physical characteristics of the final product. For example, plasticizers, incorporated to provide the required flexibility, have been known to cause threat to the health. The ink used on packaging material for printing the name, ingredient and nutritional information poses a greater threat to the safety of the product as compared to that of base packaging material. The adhesives used to seal packages also migrates to the food. Most commonly used adhesives in food packaging are pressure-sensitive, cold seal, hot melt, solvent-based, water-based, solvent-free, polyurethanes and acrylics. The selection of an adhesive is based on the nature of the product, type of packaging and other requirements, for instance the incorporation of required odor volatiles into a cold seal to improve perception of product during opening. The components on the printed surface of the packaging material migrates into the adhesive layer during storage and then move into food. Thus, the shelf life of food depends on various parameters and can be maintained by altering the composition and formulation of packaging material, its processing parameters and the environment in which the product is stored.

6.5. Rate of deteriorative reactions

Several deteriorative reactions such as chemical, microbiological and biochemical reactions occur in food. The rate of the reactions depends upon intrinsic as well as extrinsic factors. A deteriorative reaction that occurs in food during processing and storage can be controlled by quantitative study based on the kinetics of deterioration of food. Quantitative study requires a measurable index of deterioration; means, a physical, chemical or sensory measurement that may be used reproducibly to evaluate the variations that take place over time. Any change in the index of deterioration correlates with changes in the quality of food. For quantitative evaluation of quality changes, the index must be stated as a function of the prevailing conditions during processing and storage to predict the variation. Thus, the evaluation of quality losses needs a mathematical model that indicates the effect of intrinsic and extrinsic factors on the deterioration index. The equation for quality loss is as follows:

$$\frac{-dC}{d\theta} = f\left(I_i, E_j\right)$$

Where $-dC/d\theta$ is the rate of change of deterioration index C with respect to time θ; a negative sign indicates the reduction in concentration of C with time; I_i and E_j are the intrinsic and extrinsic factors ($j = 1, \ldots, n$), respectively.

The rate at which the quality of food alter during processing or storage depends on intrinsic and extrinsic factors. The loss in quality of the product is correlated with loss in a specific component of food (such as vitamin).The translation of a component or quality factor A to an end product B (such as, conversion of chlorophyll to pheophytin; or conversion of ascorbic acid to brown pigments) is written as:

A $\longrightarrow$ intermediate product $\longrightarrow$ B

The quality is directly proportional to the power of the concentration of the reactant or product:

$$\frac{-dA}{d\theta} = kA^n \quad (1)$$

$$\frac{dB}{d\theta} = kB^n \quad (2)$$

where

A and B are the concentrations of quality factor measured; θ is the time; k is the rate constant (reliant on extrinsic factors); n is a power factor also called the order of the reaction, which states the dependency of rate on concentration of A or not. The value of n can be a fraction or a whole number dA/dθ and dB/dθ is the change in concentration of A or B with time.

6.5.1. Zero-order Reaction

In pseudo-zero-order reaction, n=0 and equation (1) can be written as:

$$\frac{-dA}{d\theta} = kA \quad (3)$$

This implies that the rate of loss of A is constant with respect to time and independent of the concentration of A. Integrating the equation between concentration A_o and A at time 0 and θ respectively. After rearranging the equation becomes:

$$A_o - A = k\theta \quad (4)$$

Or

$$A_o - A_e = k\theta_s \quad (5)$$

where A_e is the value of A at the expiration of shelf life and θ_s is the shelf life in days, months, and years.

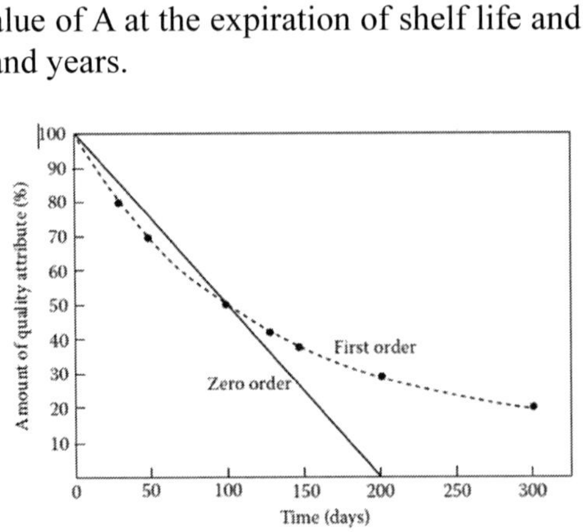

Fig. 2: Effect of order of reaction on quality with respect to time

For zero-order reaction, a graph of the quantity of A remaining versus time forms a straight line with the slope equal to the rate constant k. In other words, the loss of quality per day is constant when the extrinsic factors remains constant. Examples of zero-order deteriorative reactions are lipid oxidation and enzymatic degradation.

The shelf life can be indicated with regard to the fraction of shelf life remains

$$f_r = 1 - f_c \tag{6}$$

where f_c and f_r are shelf life consumed and remaining shelf life respectively.

Thus, for any temperature T_s;

$f_r\theta_s = (1 - f_c)\theta_s$ = shelf life at temperature T_s

Therefore, the shelf-life at any temperature is the fraction of remaining shelf life at that temperature.

6.5.2. First-order Reaction

The food that do not follow pseudo zero-order reaction deteriorate according to a pseudo first-order reaction (n=1) in which rate of loss is dependent on the amount left. For n=1, the equation (1) become:

$$\int_{A_o}^{A} \frac{dA}{A} = -k \int_{0}^{\theta} d\theta \tag{7}$$

$$ln \frac{A}{A_o} = -k\theta \tag{8}$$

$$A = A_o - e^{-k\theta} \tag{9}$$

$$\text{Or } A_e = A_o - e^{-k\theta s} \tag{10}$$

A plot of first-order reaction is shown in Figure 2. The unit of k is $time^{-1}$. Vitamin loss in canned food, microbial production of off-flavors are examples of first-order reaction.

6.5.3. Shelf Life Plots

Shelf life plots are drawn to evaluate the effect of change in temperature on the quality of the food. Various models are used to characterize the relation between the rate of a reaction and temperature. Among available models, Arrhenius and linear models are most commonly used and are shown in Figure 3.

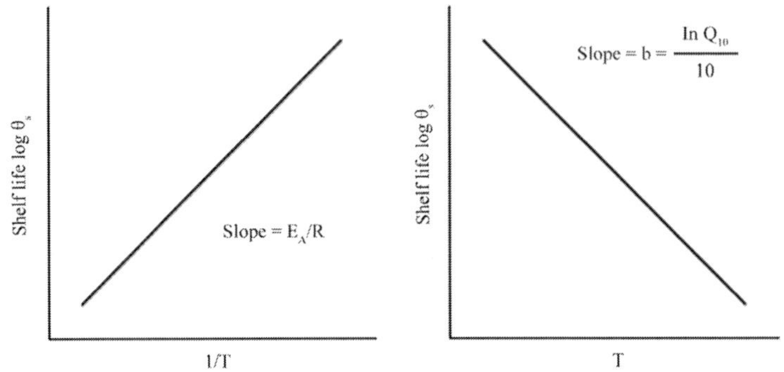

Fig. 3: (a) Arrhenius plot of shelf life log (θs) versus reciprocal of the absolute temperature and (b) linear plot of log shelf life versus temperature (°C)

The equations of these plots are as:

$$\theta_s = \theta_o exp\frac{E_A}{R}\left[\frac{1}{T_s} - \frac{1}{T_o}\right] \tag{11}$$

$$\theta_s = \theta_o e^{-b(Ts - To)} \tag{12}$$

Where, θ_s is the shelf life at temperature T_s

θ_o is the shelf life at temperature T_o

A small error is there in using simpler linear plot instead of Arrhenius plot when a temperature range less than ±20°C is used.

The deteriorative reactions in the food products are classified as zero order and first order. These reactions can be applied to forecast the level of deterioration as a function of temperature.

Zero-order prediction reaction: Equation 5 derived for change in quality factor A can be written as:

$$A_o - A_e = k_z \theta_s \tag{13}$$

where

A_e is the value of A at the end of shelf life

A_0 is the initial value of A

k_z is the zero-order rate constant ($time^{-1}$)

θ_s is the shelf life in days, months or years

For variable time–temperature storage conditions, Equation (13) can be written as follows

$A_e = A_o - \sum(k_i \theta_i)$ (14)

where $\Sigma k_i\theta_i$ is the sum of the product of the rate constant k_i, at temperature T_i, multiplied by the time interval θ_i, at the average temperature T_i, for the given time period $\Delta\theta$.

6.6. Shelf-life Determination

The properties of packaging material plays an important role in predicting the shelf life of the product. The shelf life of the product depends upon the barrier properties of the packaging material from moisture gain or O_2 uptake. The glass container and metal cans are impermeable to the odors, gases and water vapor whereas paper packing material is permeable to these. The plastic-based packaging material impart varied level of protection, depending upon the type of polymer used in the manufacturing of the package.

In steady state condition, if the pressure difference is kept constant across the polymer, then a vapor/gas will permeate at a constant rate through a thermoplastic material. The diffusive flux (J) of a gas or vapor in a thermoplastic material is defined as the amount of permeate pass through the surface of unit area normal to the direction of flow during unit time and is given as follows:

$$J = \frac{Q}{At} \quad (15)$$

where Q is the total amount of permeant which has passed through area A in time t.

As per Fick's law of diffusion equation:

$$J = -D\frac{\delta c}{\delta x} \quad (16)$$

where

J is the flux (or rate of transport) per unit area of permeant through the polymer

c is the concentration of the permeant

D is the diffusion coefficient

$\delta c/\delta x$ is the concentration gradient of the permeant across a thickness δx

D denotes the speed of permeant at which it diffuses through the polymer; it is a function of temperature and may be a function of concentration.

When a steady-state of diffusion has been reached the value of J is constant and Equation 16 can be integrated across the total thickness of the polymer X and between the two concentrations by assuming D to be constant and independent of c:

$JX = - D(c_2 - c_1)$

By substituting Equation 15, the expression can be rewritten as:

$$Q = \frac{D(c_2 - c_1)At}{X} \tag{17}$$

When the permeant is a gas, the vapor pressure p is measured which is at equilibrium with the polymer, instead of the actual concentration. Henry's law is applied at sufficiently low concentrations and c can be expressed as:

$$c = Sp \tag{18}$$

where S is the solubility coefficient of the permeant in the polymer that indicates the amount of permeant in the polymer. By combining Equations 17 and 18,

$$Q = \frac{DS(p_1 - p_2)At}{X} \tag{19}$$

The product DS is referred to as the *permeability coefficient* or *constant* and is represented as:

$P = DS$

The kinetic term (D) represents the dynamic of the penetrant-polymer system while the thermodynamic term (S) depends on the penetrant–polymer interactions.

P signifies the ease with which a gas permeates through a polymer when a pressure gradient is applied:

$$P = \frac{QX}{At(p_1 - p_2)} \tag{20}$$

Or

$$\frac{Q}{t} = \frac{P}{X} A(\Delta p) \tag{21}$$

Where

P/X is called the permeance

A is the surface area of the package

p_1 and p_2 are the partial pressure of water vapor outside and inside the package

Q/t is the rate of gas or vapor transport across the film

The moisture vapor transfer either to or from a food package is determined assuming constant P/X which means the temperature and humidity of external environment are constant. Another assumption is that the moisture gradient inside the package is negligible, means the package is resistant to water vapor transport.

The critical point regarding equation (20) is that the internal vapor pressure changes with the moisture content of the food at any time. Therefore, the rate of change in moisture is not constant but drops as Δp gets smaller. Hence, some functions of p_2 as a function of moisture content must be inserted into the equation to be able to make proper predictions. If a constant rate is assumed, then the food will be overprotected.

In low and intermediate moisture foods, the internal vapor pressure is determined solely by the moisture sorption of the food. The GAB model is preferred in describing the sorption isotherm. If linear model is used, the result can be integrated directly, but if the GAB model is used, then it must be numerically evaluated.

In the simplest case when the isotherm is treated as a linear function

$$m = ba_w + c \tag{22}$$

where

m is the moisture content in g H_2O g^{-1} solids

a_w is the water activity

b is the slope of the isotherm

c is the constant

The moisture content can be substituted for the water gain using the relationship

$$m = \frac{W\left(weight\ of\ water\ transported\right)}{W_s\left(weight\ of\ dry\ solids\ encolsed\right)} \tag{23}$$

$$\therefore W = m\ W_s \tag{24}$$

And

$\delta W = \delta m\ W_s$ (25)

By substitution

$$\frac{\delta W}{\delta t} = \frac{\delta m W_s}{\delta t} = \frac{P}{X}A + [\frac{p_0 m_e}{b} - \frac{p_0 m}{b}] \tag{26}$$

Which on rearranging, gives

$$\frac{\delta m}{m_e - m} = \frac{P}{X}\frac{A}{W_s}\frac{p_0}{b}\delta t \tag{27}$$

And integrating

$$ln\frac{m_e - m_i}{m_e - m} = \left[\frac{P}{X}\frac{A}{W_s}\frac{p_0}{b}\right]t \tag{28}$$

Where

m_e is the equilibrium moisture content of the food if exposed to external package RH

m_i is the initial moisture content of the food

m is the moisture content of the food at time t

p_o is the vapor pressure of pure water at the storage temperature (not the actual vapor pressure outside the package)

A plot of the unaccomplished moisture change (the term on the left side of Equation 28) versus time is a straight line with a slope equivalent to the bracketed term on the right-hand side of the equation.

The end of product shelf life is reached when $m = m_c$, the critical moisture content, at which time t = θ, the shelf life. Thus, Equation 28 can be rewritten as

$$ln\frac{m_e - m_i}{m_e - m_c} = \frac{P}{X}\frac{A}{W_s}\frac{p_0}{b}\theta_s \tag{29}$$

The relationship between initial, critical and equilibrium moisture content is illustrated in Figure 4

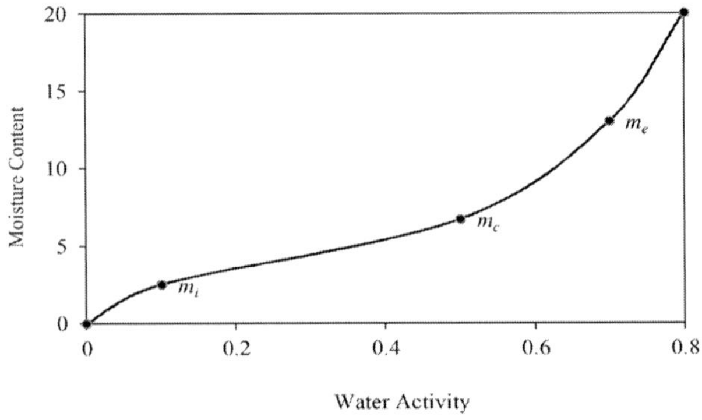

Fig. 4: Typical moisture sorption isotherm

The equation 29 is found to be useful in determining the effect of variation in the external condition (such as temperature; humidity), surface area to volume ratio of the package and changes in initial moisture content of the food product.

Example 1: The cookies has an initial moisture content m_i of 5%. The critical moisture content m_c of 6.9% due to loss of crispiness of the product. The equilibrium moisture content m_e at 25°C is 8.4%. The slope of line is 0.07 g H_2O/g solids/unit a_w. Calculate the shelf life of the cookies packaged in (a) polyethylene with permeability 0.05 g H_2O/day.m^2.mmHg and (b) aluminum foil with permeability 0.006 g H_2O/day.m^2.mmHg. The weight of dry cookies in the package is 24 g and the dimension of packaging bags are 12 cm × 20 cm. The package is stored at 25°C and 75% RH and the vapor pressure of pure water at storage temperature is 28.349 mmHg.

Solution:

Given: m_i = 5%; m_c = 6.9%; m_e = 8.4%

Slope (b) = 0.07 g H_2O/g solids/unit a_w

Vapor pressure of pure water at storage temperature =28.349 mmHg

Weight of dry cookies in the package = 24 g

Surface area of bag = 12 × 20 = 240 cm^2 = 0.024 m^2

(a) For polythene,

$$\frac{P}{X} = 0.05 \text{ g } H_2O/\text{day.m}^2.\text{mmHg}$$

Equation 29 can be used:

$$ln\frac{m_e - m_i}{m_e - m_c} = \frac{P}{X}\frac{A}{W_s}\frac{p_o}{b}\theta_s$$

Substituting into Equation 29

$$ln\frac{8.4 - 5}{8.4 - 6.9} = 0.05\frac{0.024}{24}\frac{28.349}{0.07}\theta_s$$

Solving for shelf life θ_s,

$$\theta_s = \frac{ln2.267}{2.024 \times 10^{-2}}$$

= 40 days

(b) For aluminum foil,

$$\frac{P}{X} = 0.006 \text{ g } H_2O/\text{day.m}^2.\text{mmHg}$$

Substituting into Equation 29

$$ln\frac{8.4-5}{8.4-6.9}=0.006\frac{0.024}{24}\frac{28.349}{0.07}\theta_s$$

$$\theta_s=\frac{ln2.267}{2.429\times10^{-3}}$$

$$=337 \text{ days}$$

The shelf life is inversely related to the water vapor permeances of the packaging material; since P/X for polythene is 8.3 time higher than that of aluminum, the shelf life of later is 8.4 times higher than former.

6.7. Microbial Shelf-life Prediction

The microbial spoilage of the food is a major concern for manufacturers, retailer and consumers. The microbial shelf life is examined by the growth of pathogenic or spoilage microorganisms. In the traditional method, the microbial shelf life of the food for spoilage microorganisms is determined by storing the food at different temperature and spoilage is evaluated by sensory method or microbial count. In the case of pathogenic microorganisms, the traditional methods have been challenging due to organism of concern followed by storage at different temperatures and microbial analysis at intervals. The inoculated packs are used for the process where elimination of a particular microorganism is required such as heat treatment in canning. In the microbial shelf-life prediction of food, the first step is to detect the specific spoilage organisms (SSOs). SSOs are the fraction of the total microflora that cause the spoilage of food under particular environmental conditions. A certain level of deterioration by SSOs mark the end of shelf-life. After identifying the SSOs and the extent of environmental conditions over which a particular SSO cause spoilage, the further step is to determine the SSO population at which spoilage results the end of shelf-life. The microbial count in assessing the shelf-life of a food material varies with the type of food, packaging and storage conditions. The SSO counts of 10^6–10^8 CFU per gram are usuallyconsidered as an upper limit of the quality. Pathogenic bacteria for instance, *Bacillus cereus* and *Staphylococcus aureus* with a limit of 10^5 CFU per gram has been used for risk management in the food supply system.The time required for the growth of pathogens to a set limit is the minimum requirement to control the shelf-life. There should always be a safety margin that results in shorter shelf life that depends on the initial level of contamination.

During the exponential/logarithmic growth phase, microbes follows first order reaction. The equation 8 can be written as

$$ln\frac{A}{A_o} = \mu_{max}\theta_{doub} \quad (30)$$

where

A_o is the initial number of microorganisms when $\theta = 0$

$A = 2A_o$ (number of organisms has doubled)

μ_{max} is the maximum specific growth rate constant (analogous to k in chemical reactions), θ_{doub} is the time for number of organisms to double (i.e. the generation or doubling time). On substituting $2A_o$ for A,

$$\ln 2 = \mu_{max}\ \theta_{doub} \quad (31)$$

$$\mu_{max} = \frac{0.693}{\mu_{max}} \quad (32)$$

In shelf life studies, the time to attain a critical upper limit of cells is of great interest and for this the equation 15 can be written to include a lag time θ_{lag} as

$$\theta_s = \theta_{lag} + \frac{1}{\mu_{max}} ln\frac{N_c}{N_o} \quad (33)$$

where

N_o is the initial number of microorganisms at $\theta = 0$

N_c is the critical or maximum permitted number of microorganisms

θ_s is the time for number of microorganisms to reach critical level (i.e., microbial shelf life)

The parameter log N_o is evaluated by the initial level of contamination of the food product and depends on the quality of raw material and food processing conditions.The other parameter log N_c denotes the critical or maximum density of cell that is attainable in given conditions. Lag time (θ_{lag}) and maximum specific growth rate (μ_{max}) are dependent on environmental conditions and directly affect the time required to reach a critical level of microbial density in relation to acceptable quality. Therefore, these two parameters are mostly used in determining the effect of packaging conditions on microbial shelf life.

The time (θ_s) to reach a critical limit of cell density of N_c is the shelf life and can be calculated using

$$\theta_s = \theta_{lag} + \frac{1}{\mu_{max}} ln\frac{N_c}{N_o} \quad (34)$$

Example 2: Beef pie packaged in plastic film is stored at 8°C. The initial level of contamination of soup by *Clostridium perfringens* after packaging is 10^2 CFU/g, and maximum permitted level of microorganisms is 10^4. A generation or doubling time of *Clostridium perfringens* at 8°C is 3 h, calculate the time for which the soup can be stored before the maximum permissible level of microorganism is reached assuming no time lag.

From equation 32

$$\mu_{max} = \frac{0.693}{3} = 0.231 \text{ h}^{-1}$$

Substituting in equation 34

$$\mu_{max} = \frac{0.693}{3} = \frac{1}{0.231} ln \frac{10^4}{10^2} = 19.9 \text{ h}$$

Beef pie can be stored for 19 h at 8°C. If this shelf life is not sufficient at given temperature, the storage temperature could be reduced.

6.8. Accelerated shelf-life Testing

The principle of chemical kinetics has been applied to develop accelerated shelf-life testing (ASLT) in which the effect of extrinsic factors such as humidity, gas atmosphere, temperature, and light is quantified on the rate of deteriorative reactions. The food is placed in a controlled environment in which any one or more of the extrinsic factors is kept at a higher level than usual. In such condition, the rate of deteriorative reaction becomes faster than the normal rate that results in a reduction in the shelf life of the product. The magnitude of the accelerated reactions can be evaluated and the "true" shelf life of the food in normal conditions is determined. The evaluation of shelf life that would take a year can be completed within 1-2 months if the storage temperature is increased to 40°C from 20°C.

Most of the food products have a shelf life of at least one year and determination of shelf life by changing the product formulation (e.g., addition of thickener or new antioxidant), the process (e.g., time/temperature combination) or the packaging (e.g., latest polymeric film) requires the shelf-life trials to continue for the time period equivalent to the desired shelf life of the food material. The manufacturers can not wait for such a long period to know that the new product, process or packaging will provide an adequate shelf life or not. In ASLT, the time required to evaluate the shelf life is reduced. But the use of ASTL is limited to very simple chemical reactions. The food technologist has to ratify the ASLT results for a specific food by performing shelf-life tests under definite environmental conditions. When a relationship between actual shelf life and ASLT has been formed for a certain food, then ASLT can be used for that food if any of the process or package parameters are to be examined.

As discussed earlier, the quality loss for most of the foods follows a zero-order or first-order reaction. At a given level of deterioration and reaction order, the rate constant is inversely proportional to the time to reach certain amount of quality loss. The Q value is a quotient representing how rapidly the reaction proceeds at temperature T_2 as compared to lower temperature T_1. Thus, by taking the ratio of the shelf life between any two temperatures with difference of 10°C, the Q_{10} of the reaction can be determined as follows:

$$Q_{10} = \frac{k_{T+10}}{k_T} = \frac{\theta_{sT}}{\theta_{sT+10}} \tag{35}$$

where

θ_s is the shelf life at temperature T °C

θ_{sT+10} is the shelf life at temperature (T+10) °C

6.9. ASLT Procedure

The procedure involves the following steps:

(i) Determine the key critical descriptor(s) or IoFs that will lead to the deterioration of quality of the food, and accordingly decide the tests (sensory and/or instrumental) that should be conducted on the food product during the trial.

(ii) Select the package for analyses. It is recommended to select the most cost-effective material as a wide range of packaging material are tested.

(iii) Choose the extrinsic factors that are to be accelerated. The storage temperatures used for ASLT testing is as follows:

Product	**Test Temperatures (°C)**	**Control (°C)**
Frozen	−7, −11, −15	<−40
Chilled	5, 10, 15, 20	0
Dry and IMF	25, 30, 35, 40, 45	-18
Canned	25, 30, 35, 40	4

Plot given in Fig 4 is used to determine the time for which product under test must be held at each test temperature. If no Q_{10} values are known, then an open-ended ASLT have to be performed using a minimum of three test temperatures.

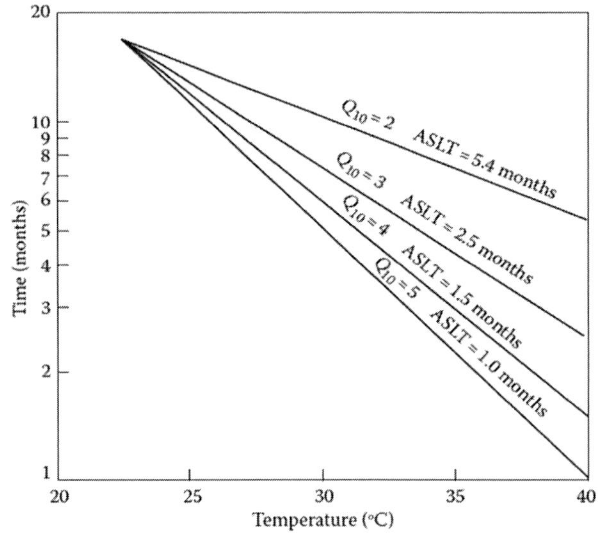

Fig. 4: Hypothetical shelf life curve for several Q10's passing through a shelf life of 18 months at 23°C. Accelerated shelf life times (ASLT) are those required at 40°C for various Q10's. (Labuza & Kamman, 1983.)

(iv) Determine the frequency of the tests. A general criterion is that the time interval between tests at given temperature below the highest temperature should not be more than

$$f_2 = f_1 Q_{10}^{\Delta T/10} \tag{36}$$

where

f_1 is the time between tests in days, weeks at the highest test temperature T_1

f_2 is the time between tests at any lower temperature T_2

ΔT is the difference between T_1 and T_2

(v) Estimate the number of samples that is to be kept at each test condition including samples under control.

(vi) Commence the ASLTs and plot the data as if required, the frequency of sampling can be altered (i.e. decreased or increased).

(vii) From each test, estimate the value of k or θ_s and draw suitable shelf life plots to assess the potential shelf life of the food material under normal storage conditions. The shelf life plots should specify the shelf life of the product at least as long as anticipated by the company, then only the product has a chance to perform satisfactorily in the market place.

References

Burdurlu, H. S., Koca, N., & Karadeniz, F. (2006). Degradation of vitamin C in citrus juice concentrates during storage. Journal of Food Engineering, 74(2), 211-216.

Kilcast, D., & Subramaniam, P. (Eds.). (2000). The stability and shelf-life of food. Woodhead Publishing Ltd., Cambridge

Mauer, L. J., Ozen, B.F. (2004). Food Packaging. In: Smith, J.S. and Hui, Y.H. (Ed.), Food Processing Principles and Applications (pp. 101-132). Blackwell Publishing, 2121 State Avenue, Ames, Iowa 50014, USA

Man, C. D., & Jones, A. A. (Eds.). (1994). Shelf life evaluation of foods. Glasgow: Blackie Academic & Professional.

Phimolsiripol, Y., & Suppakul, P. (2016). Techniques in shelf life evaluation of food products. Reference Module in Food Science, Elsevier. https://doi.org/10.1016/B978-0-08-100596-5.03293-5

Robertson, G.L. (Ed.) (2010). Food Packaging and Shelf Life: A Practical Guide. CRC Press, Boca Raton, Florida, USA.

Robertson, G.L. (Ed.) (2013). Food Packaging Principles and Practice Third Edition. CRC Press, Boca Raton, Florida, USA.

Sothornvit, R., & Kiatchanapaibul, P. (2009). Quality and shelf-life of washed fresh-cut asparagus in modified atmosphere packaging. LWT-Food Science and Technology, 42(9), 1484-1490.

Steele, R. (Ed.). (2004). Understanding and measuring the shelf-life of food. Woodhead Publishing.

Van Boekel, M. A. (2008). Kinetic modeling of food quality: a critical review. Comprehensive Reviews in Food Science and Food Safety, 7(1), 144-158.

7

Aseptic Packaging

Mayank Kumar Nigam[1] and Kshitiz Kumar[2]

[1]*Department of Food Engineering, NIFTEM, Sonipat, Haryana, India*
[2]*A D Patel Institute of Technology, CVM University, V V Nagar, Gujarat, India*

7.1. Introduction

In the modern lifestyle, people are extremely busy in their work and lack the time to cook their own foods. They mainly depend upon processed and packaged food for their nutritional intake. So it becomes essential that the packaged products should be safe, hygienic and shelf stable. To preserve the food for long duration, thermal processing is most suitable method. Aseptic packaging is a form of thermal food processing. Aseptic packaging can be defined as the filling of sterile containers with a commercially sterile product under sterile conditions and then sealing the containers so that reinfection is prevented; that is, they are hermetically sealed. The word "aseptic" is derived from the Greek word "*septicos*" meaning absence or exclusion of any spoilage or unwanted microorganism from product, package or other specific areas. In simple term commercial sterility means the absence of microorganisms capable of growing in the food at non-refrigerated conditions at which the food is likely to be held during manufacture, distribution and storage. The word 'hermetically' means strictly air tight thus preventing the movement of microorganism, gas or water vapour between the packaged product and the outside environment. Processing and packaging of food material aseptically is one the best techniques especially for those foods which have less pH (>4.5) such as juice (including both fruits and vegetable), milk, dairy products, pudding, soups, sauces.

Aseptic packaging is totally different from the traditional thermal processing like canning. Traditional method of packaging involves first filling of the food products in the container followed by thermal processing whereas aseptic packaging of foods consist of filling sterilized and cooled food into presterilized containers, followed by hermetic sealing in aseptic work zone (Figure 7.1).

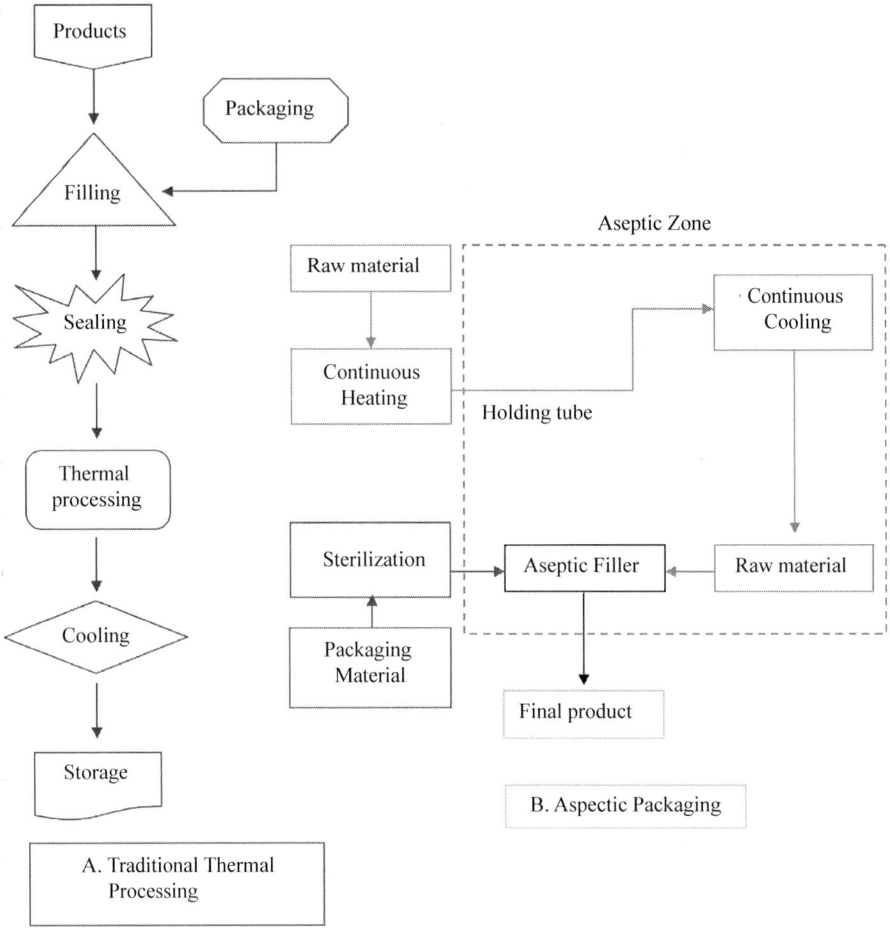

Fig. 1: Comparison between traditional thermal processing and aseptic packaging

7.2. Reasons for Aseptic Packaging

The three major reasons for the aseptic packaging usage are as follows:

(i) To take advantage of high temperature short time (HTST) sterilization processes which gives product of superior quality than low temperature long time (LTLT) process.

(ii) To utilize such containers/materials for packaging purpose which cannot endure high temperature encountered in traditional thermal processing and are unfit for in package sterilization.

(iii) To extend the shelf life of packaged products at ambient temperatures (eliminating the need of refrigerated condition of storage).

7.3. Application of Aseptic Packaging

Aseptic packaging is commercially used for:

(i) Packaging of pre-sterilized and sterile product packaging. Here food products are sterilized separately and then packed in pre-sterilized packaging material in sterile environment. Example include milk and milk product, puddings, desserts, fruit and vegetable juices, soups, sauces.

(ii) Packaging non-sterile materials to prevent microbial contamination. Here food products need not be sterilized as they are preserved by other method of preservation such as lowering of pH, use of preservatives. Examples include yogurt and other fermented dairy products.

7.4. Aseptic Packaging Procedure and Requirements

Aseptic packaging involves various steps (Figure 1) which must be followed to get quality products. The food product to be packed in aseptic packaging system are sterilized by HTST or UHT procedure. The packaging material is sterilized separately from the food material. The pre-sterilized food products are filled and sealed in the pre-sterilized packaging system in the sterile environment. Some of the basic requirements of aseptic packaging are as follows:

- The container and the closure should be suitable for aseptic packaging. Container should not allow entry of microorganism in the sealed container during storage.
- Before the start of full operation, ensure the proper sterilization of all the equipments which are essential for aseptic processing such as holding tube through which movement of food product occurs as well as filler and other relevant equipment zones. The premise of production area should be regularly disinfected to eliminate contamination factors.
- Complete sterilization of all the entering media such as air, gases, and water for cleaning, heating and pneumatic applications.
- The container or the packaging material must be sterilized after it is formed and before being filled. If the container is sterilized in single step, it should be carried out in aseptic zone. For two stage sterilization procedure, the second stage should form part of the filling operation.
- Adequate sterilization of food product. Before filling the final products in the package, it must be properly sterilized followed by cooling.

- During filling and sealing sterile environment in the whole processing area must be maintained.
- If any closure is needed it must be sterilized immediately before it is applied.

7.5. Basics of Sterilization

The process of sterilization eliminates the growth of microorganism and makes the food products and packaging materials completely sterile and free from microbial load. After sterilization products are free from colony forming unit (CFU) and it can only be contaminated when the products are reintroduced to the contaminated or infected environment. For aseptic processing high temperature short time (HTST) or ultra high temperature (UHT) is used for sterilization process which also ensures the preservation of nutritional and sensory characteristics of food products. UHT treatment is performed within the temperature range of 130-150°C for 2-8 second. Product having low viscosity like milk is treated at upper range of the temperature while products having high viscosity treated at lower range of temperature. Similarly, autoclaving of any product at 121°C for 20 min at 15 psi guarantee the sterility of that particular product. This results into destruction of CFU but at the same time it also affects the nutritional and sensory properties of food products. Thus to preserve these properties besides achieving commercially sterile food products, HTST and especially UHT is used in aseptic packaging. In such case, the reduction of microorganism follows a logarithmic function as shown in equation 1.0.

$$SV = \log_{10}\left(\frac{N_0}{N_t}\right) = \frac{\mathrm{F}}{\mathrm{D}} \tag{1.0}$$

Where SV is known as sterilizing value, N_0 is initial microbial count at time t=0 and N_t is final microbial count at time t=t. F is the thermal death time (TDT) and D is decimal reduction time (D value). D-value is the time in minutes at any temperature that results in 90% destruction of the existing microbial population.

The lethal effect of any sterilization process can be obtained by summation of lethality obtained at each temperature for unit time known as F value. The most usual form of F-value equation is

$$F = \int 10^{(T-T_{ref})/Z}\, dt \tag{2.0}$$

where T is temperature, T_{ref} is reference temperature (usually 121.1 °C), Z is the thermal resistance of microorganism, t is the time. Z value is defined as the increase in temperature necessary to cause 90% reduction in decimal reduction time (D-value). In similar lines, Z value is also defined for nutrients and quality parameters where it can be considered as increase in temperature necessary to cause 90% reduction in specific nutrients or quality parameters (enzymes, protein, vitamins, colour, texture).

The advantages of using HTST and UHT process can be clearly understood by comparing the Z value for microbial destruction to Z values for the effect of heat on nutrients. The Z value for the microbial destruction is 7-12ºC while Z value for nutrients vary from approximately 25-50 °C. This means that for every 10 °C rise in temperature microbial inactivation increases tenfold while nutrient quality loss is only twofold. Cooking index/value (C_0) or C-value indicates the overall deterioration of nutrient or sensory quality during the heat treatment. It is similar to the F_0 value except the reference temperature is 100ºC unlike 121.1ºC for microbial inactivation. In simplest form cook value can be defined by the equation

$$C = \int 10^{\left(T-T_{ref}\right)/Z_c}\, dt \qquad (3.0)$$

From equation 2.0 and 3.0 it is clear that the high temperature has less impact on nutrient and quality deterioration because of increased Z value.

This can be understood form following example. For a UHT process, T=140 °C, T_{ref}= 121.1°C, Z = 10 °C (for bacterial spore) and Z_c = 40 °C (for colour degradation). The F value for 2 sec of holding can be obtained from modified form of Equation 2.0 as:

$$F = 10^{\left(T\text{-}T_{ref}\right)/Z} \times t$$

$$F = 10^{140\text{-}121.1/10} \quad 0.033\text{=}2.6$$

The C value for same process for colour as obtained from modified form of Equation 3.0 as

$$C = 10^{\left(T-T_{ref}\right)/Z_c} \times t$$

$$C = 10^{140-100/40} \times 0.033 = 0.33$$

In above UHT process (140 °C for 2 sec) effect of thermal processing at is almost 8 times more lethal on microbial inactivation than colour degradation.

Now for a low temperature long time (LTLT) process say T=63 °C, T_{ref} = 121.1°C, Z= 10 °C (for bacterial spore) and Z_c= 40 °C (for colour degradation). The F value for 30 min of holding from equation 2 will be

$$F\text{=}10^{63-121.1/10}\times 30\text{=}2.6\text{=}4.6\times 10^{-5}$$

The C value for the process for colour as obtained from equation 3 is now

$C=10^{63-100/40}\times 30=3.56$

In above LTLT process (63 °C for 30 min) effect of thermal processing at is almost 76000 times more lethal on colour degradation than microbial inactivation.

Thus, HTST and UHT are highly effective for the prevention of microorganism and have very less impact on quality of products. One problem associated with UHT processing is that the enzymes like proteases, lipases, and peroxidases do not gets inactivated due to their heat resistance properties which impact the quality of food product like off flavor and age gelation.

7.6. Product Processing/Sterilization

For aseptic processing, the process of UHT commonly involves four main steps for pre-sterilization of the food products.

(i) Preheating of the food products at a defined temperature.

(ii) Final heating at sterilization temperature.

(iii) Holding the temperature for a definite time.

(iv) Instant cooling at ambient temperature to prohibit damage of products.

While carrying out product processing, the processing steps should be shorter to avoid loss of quality due to high temperature. The sterilization methods of food products during aseptic processing can be classified into two i.e. indirect heating approach and direct heating approach. In indirect heat exchange products and heating medium are kept separately and the exchange of heat occurs through heat conducting surface. While in direct heat exchange method, heating media such as steam comes in direct contact with the food products at required temperature to sterilize the product followed by pre-cooling and aseptically filling into the packets. Indirect heat exchange includes plate heat exchanger (PHE), tubular heat exchanger and scraped surface heat exchanger. Direct heat exchange methods include techniques like steam infusion where products is inject into the steam flow and steam injection where steam is injected into the food products.

Both the system of heating i.e. direct and indirect heating have their own advantages and limitations. Direct method of heating requires very less time as compared to the indirect method. It takes less than one second to attain the sterilization temperature in direct heating. The undesirable flavor compounds produced during heating are also removed in direct heating system. However, steam injection in direct heating induces aggregation of proteins which gives stringent mouth feel. The cost efficiency of indirect heating method is high because of low energy consumption and lower installation and maintenance expenditure. Apart from this, indirect method heating equipments are more automatic so they are easier to handle and thus used by most of the industries.

Importance of aseptic packaging has grown tremendously in last few decades and various new technology for the product sterilization have emerged. The newly developed methods do not require heat to kill microorganism rather it uses methods like pressure, radiation, electricity etc. for cellular death for microorganisms. It includes radio frequency heating, induction heating, dielectric heating, microwave heating, Ohmic heating, pulsed electric fields (PEFs), ultraviolet (UV)-radiation, pulsed light, ultra high pressure (UHP). Techniques like PEF and UHP are already in use and have been acknowledged for the efficiency and product quality. But high cost of operation is one of limitation of its wider use.

The aseptic processing for the different food products can be classified into two categories:

7.6.1. Low Acidic Product (pH>4.5)

The method of treating low acid (pH≥5.3) and medium acid food (pH 5.3-4.5) in aseptic packaging is same as the principle of canning of food products. In low acid food such as meat, fish, milk, pasta and most vegetables, pathogenic microorganism such as *Clostridium botulinum* can grow. Thus, in this case aseptic processing must be designed to achieve 12-D reduction of *Clostridium botulinum* spores. This not only eliminates the heat resistant bacteria but also ensures the killing of other microorganism at a faster rate. The packaging material used for low acidic product must have same level of sterility as the food products. Since in low acidic food products pH is not an able barrier to pathogenic microbial growth, therefore low acidic product packaging material must have stricter sterility control than acidic food. Some examples of sterilization techniques for the packaging material include chemical technique like passing through hydrogen peroxide bath of 30% concentration and heated to 70 °C for 5-6 seconds or physical techniques like sterilization by saturated steam or superheated steam. For low acid food products packaging, *Bacillus subtilis* is the most resistant organism to H_2O_2 (Hydrogen Peroxide) sterilization and if 4D reduction of *B. subtilis* is achieved in packaging then required reduction for *C. botulinum* will automatically be achieved.

7.6.2. Acidic Products (pH<4.5)

In acid (pH 4.5-3.7) and high acid (pH≤3.7) food products, for example citrus fruit juice and tomatoes products, pH act as barrier to pathogenic microbial growth. Such food products are more vulnerable to spoilage microorganism than pathogenic. Spoilage organisms are much more sensitive to hydrogen peroxide sterilization of packages. Unlike low acidic food product and food packaging material where *Clostridium botulinum* is the main target microorganism, here in high acid products and packaging, *Aspergillus niger* is the target microorganism for dry and moist heat sterilization of packaging materials.

7.7. Sterilization of Packaging Material

7.7.1. Required Count Reduction

The type of product, the target shelf life, and the expected storage temperature all affect the count reduction (number of D values) needed to sterilize the food contact packaging material surface. There must be a minimum of four decimal decreases in bacterial spores in nonsterile acidic goods with pH 4.5. A six decimal reduction is necessary for sterile, neutral, low acid products with pH values greater than 4.5. However, a full 12 decimal reduction process should be applied if *Clostridium botulinum* could potentially develop in the product. The number of faulty or non-sterile packages as a percentage of all the packages processed over a specific time period is known as the non-sterility rate, or error rate, E_r.

This equation can be used to calculate it:

$$E_r = A\text{-}N/R$$

where, 'N' is the quantity of microorganisms on the packaging material's food contact surface, 'A' is the area of the packaging material's food contact surface, 'R' indicates how many decimal reductions were made during the sterilizing process. To achieve a specific non-sterility rate E_r, smaller containers with a smaller food contact surface area will have proportionally less initial contamination and require a less rigorous sterilization process. In contrast, a more rigorous sterilization procedure will be needed for larger containers.

There are number of techniques which are used for sterilization of packaging materials, which may be classified into three groups:

a) Sterilization by Heat

b) Sterilization by Chemicals

c) Sterilization by Radiation

7.7.2. Sterilization by Heat

It is mainly performed either through moist heat/steam or dry heat. Moist heat/ steam is the pure gaseous form of water without any air or gases while dry heat is just opposite i.e. hot air without any water molecules. Sterilization by moist heat is done at temperature ranging from 121-129 °C along with pressure while sterilization by dry heat is done at temperature ranging from 176-232 °C for longer time. Sterilization by moist heat is substantially more effective than dry heat and is dependent on temperature and duration. In terms of effectiveness, dry heat sterilization at 170 °C for 60 minutes is similar to moist heat sterilization at 121°C for 20 minutes. High temperature for longer time (HTLT) is mainly used for metallic packaging or plastic laminates which may sustain high heat. This process requires the use of sterile air which is sterilized at temperature between 260-320 °C. The main benefits of this techniques is that it is simple, effective, easy to control and cost effective. But this techniques is limited to those packaging materials only which have high heat tolerance capacity. This method is also compatible with multilayer laminate packaging. The heat that is required to melt the raw materials is high enough to guarantee the sterilization of the material. To prevent the recontamination, the final forming is carried out in aseptic packaging machines.

7.7.2.1. Saturated Steam

When heat is applied to water at specific pressure, the temperature rises continuously until the temperature at which it reaches its boiling point. As more heat is applied, the water vaporizes and transforms into steam. Saturated steam is defined as steam that exists at the same temperature as the water from which it originates. So, when water is heated (sensible heating) to the boiling point and then evaporated with additional heat (latent heating), saturated steam is produced.

Steam is the most dependable sterilant. However, the steam (and consequently the packaging material with which it comes into contact) must be under pressure in order to reach temperatures high enough to complete sterilization in a matter of seconds, necessitating the employment of a pressure chamber. In order to prevent the pressure chamber from interfering with the passage of heat from the steam to the package surface, any air that enters into it with the packaging material must also be eliminated. Last but not least, condensate is created when steam condenses on the heated surface of the packaging material. If not removed, condensate could stay in the container and dilute the product. Despite these issues, sterilizing plastic containers with saturated steam under pressure is still practiced. For instance, after deep drawing, molded PS cups and foil lids are steam-treated for 1.4 and 1.8 seconds, respectively at 165°C and 600 kPa (lids). The outside of the cups is simultaneously cooled in order to keep the heating effect to the inner surface. It has been demonstrated that using this method can reduce *Bacillus subtilis* spores by five to six decimal places. Moist heat is not suitable for sterilization of paper based packaging material as it can cause blistering and delamination. It also impairs the heat sealing characteristics of the plastic. It is suitable for sterilization of metal containers and equipments used in aseptic processing such as hold tank, homogenizers and fillers.

7.7.2.2. Superheated Steam and Hot Air

The steam becomes superheated steam if it is heated (sensible heating) beyond the saturation point. Superheated steam and hot air deliver dry heat. As dry heat is less effective than moist heat, higher temperature is required for sterilization by superheated steam or hot air. The maintenance of high temperatures ensures asepticity by eliminating all microorganism. Since 1950s, the Martin-Dole aseptic canning process is used to sterilize tinplate, aluminum cans and lids using superheated steam. Depending on the type of construction material used, the cans are continually heated to 220°C to 256°C superheated steam at normal pressure for 36–45 s. Aluminum cans has a shorter heating time due to their higher thermal conductivity. For paper based packaging material hot air is preferred over superheated steam. Both these sterilant does not affect the quality of packaging material. High temperatures of headspace gas during sealing reduces the level of oxygen and diminishes the chance of deterioration due to oxidation reactions. Plastic film are sensitive to high temperature and hence not sterilized by high temperature dry heat sterilants.

7.7.3. Sterilization by Chemicals

Various types of chemical in liquid and gaseous form are used for the sterilization of packaging materials. The most commonly used chemical for the process of sterilization are hydrogen peroxide (H_2O_2), ethylene oxide and peracetic acids (PAA).

7.7.3.1. Hydrogen peroxide

For sterilization of packaging material, H_2O_2 is used at concentration ranging from 30-35%. After dipping in H_2O_2 packaging material is treated with hot air at 60-125 °C. Hot air supplements the lethal effect of H_2O_2 and removes any residual H_2O_2 from the surface. Application of H_2O_2 at 30% concentration for 1 minute at ambient temperature resulted in only 2.5 decimal reduction in mold spore and negligible effect on bacterial spore. Hence, a minimum temperature of 70 °C for 6 s at 30% H_2O_2 is required for the destruction of resistant spores by H_2O_2. Although the exact mechanism of sterilizing effect of H_2O_2 is debatable, it has been suggested that nascent oxygen produced due to breakdown of peroxide reacts with microorganism cell metabolites leading to their death. Hydrogen peroxide can kill wide variety of microorganism including heat resistant spores. It is popular sterilizing agent as it is non-toxic and decomposes to water and oxygen on heating which do not leave any taint on food. Any residual H_2O_2 below 0.25 ppm does not have any harmful effect when consumed. H_2O_2 is approved for sterilizing of aseptic packaging system used for commercial packaging of low acid food. H_2O_2 is not recommended for sterilizing packaging material used for packaging of food containing ascorbic acid due to degradation of ascorbic acid in presence of H_2O_2.

The sterilizing effect of H_2O_2 can be increased by increasing its concentration and temperature (upto 85-90 °C). D-value of spores *Bacillus stearothermophilus* and percent concentration of H_2O_2 are related as:

$$\log D = 17 - 1.3 \times \log (\%\, H_2O_2)$$

Where D is Decimal reduction time in minutes.

H_2O_2 can be applied over the packaging material in three different ways:

- ***Dipping***: In this techniques packaging materials from the reels are dipped into a bath tub containing 30-33% H_2O_2 solution. Wetting agents are added to ensure proper wetting of the surface. After dipping, packaging materials are removed from the bath and then excess solution is removed by squeeze roll. The surface is then subjected to heat either by jet of hot air or radiant heating element.
- ***Spraying/aerosol***: This techniques involves spraying of H_2O_2 on prefabricated containers and packaging system in the form of small dispersed droplets. In conventional method of spraying, a measured amount of H_2O_2 is taken and delivered through nozzle on each container. In low acid food the requirement of high microbial lethality to attain commercial sterility makes it compulsory to spray large volume of H_2O_2. This may result in longer drying time as well as chance of residual H_2O_2 beyond the permissible limit of 0.5 ppm. Decrease in droplet size from 30 μm in conventional spray to 3 μm in ultrasonic based nozzle spray increases the surface coverage from 30% to 60% and also allows less volume of H_2O_2 solution to be applied. Now the old method of spraying is replaced by use of a mixture of hot air and vaporized peroxide. Hot air having temperature around 130 °C mixed with H_2O_2 vapour and then the process of sterilization is carried out by spraying. Vapour treated surface are then flushed with low temperature sterile air to remove the

H_2O_2 residues. Application of H_2O_2 in vapour form is cost effective as less amount of H_2O_2 is required to cover a relatively large surface area.

- ***Rinsing***: The process of sterilization by rising is very much similar to the dipping method but this techniques is more appropriate for prefabricated packaged having an intricate shape. In this method the containers are drained with H_2O_2 as whole or mixture of H_2O_2 and PAA, allowed to drip dry and then completely dried using hot sterile air. This is suitable for glass containers, metal cans and blow molded plastic bottles.

7.7.3.2. Ethylene oxide

It is used to sterilize paperboard cartoon and plastic packaging material. Ethylene oxide is used in mixture with inert gas such as dichlorodifluromethane because vapour form of pure ethylene oxide is flammable and explosive. It can be used in bulk sterilization of preformed packaging material due to its high permeation. It can kill bacteria yeast and viruses by alkylation of nucleic acid and other cellular component. It has slow rate of action on the microorganism. Ethylene oxide concentration of 700 mgl^{-1} at 54 °C and 40-50 % relative humidity requires 3-4 h for sterilization.

7.7.3.3. Peracetic acids (PAA)

It is liquid sterilant effective against spores of aerobic and anaerobic bacteria. It is effective at lower temperature (20 °C) than H_2O_2. It is used in solution that contains H_2O_2 since it is is produced by the oxidation of acetic acid by H_2O_2. A 1% solution of PAA will eliminate 10^7 to 10^8 of most resistant spore strains in 5 minutes at 20 °C and the most resistant strains in 60 minutes. The maximum temperature used in PAA use is 40 °C which reduces the sterilization time by 5 times. PAA can corrode the metals such as iron, copper and aluminium. It is suitable for sterilizing packaging material such as PET.

7.7.4. Sterilization by Radiation

This method involves neither any heat nor chemicals for the process of sterilization of packaging materials. Therefore, it offers packaging materials that are free from any chemicals. It also allows the use of less heat tolerant packaging materials. This method involves use of infrared, ultraviolet, gamma rays and electron beam. These irradiation techniques does not leave any residue on the packaging material as well as on environment. Irradiation sterilizes the packaging material in few seconds only. International Atomic Energy Agency (IAEA) has recommended use of radiation as sterilizing agent for polyethylene, polyester, polypropylene, nylon, PVC etc.

7.7.4.1. Ionizing radiation

It includes the use of α-rays, X-rays, β-rays, electron beam and γ-rays. γ radiation from cobalt-60 or cesium-137 has been used to sterilize interior of empty container which cannot sustain high temperature or cannot be sterilized by other means due to their intricate shape. A radiation dose of 2.5 Mrad is enough to ensure sterility of packaging material. A dosage of 4.7 Mrad is considered as 12D reduction for *Clostridium botulinum*. γ irradiation has been used to sterilize plastic bags used in aseptic bag in box system. Owing to low consumer acceptability of γ irradiated packaging materials, a new techniques of radiation i.e. electron beam are more in use due to better penetration, high sterilization impacts and larger acceptance. A dose of 5-7 kGy of electron beam is sufficient to kill most of yeast, mold and spores. The shielding requirement for electron beam is less than the γ-irradiation because they have less penetrating power.

7.7.4.2. Pulsed Light

Electromagnetic radiation from the ultraviolet region to infrared red region can be used for sterilization. Light when applied in short pulses can sterilize the packaging material more effectively than continuous form. For sterilization the packaging material must be exposed to one light pulse of energy density 0.01-50 Jcm^{-2}. The packaging material are generally exposed to 1-20 pulses of light intensity short duration light. The pulse duration ranges from 1 μs to 0.1 s. In most cases, high levels of microbial inactivation are achieved with a few flashes applied in a split second. This system has undergone a successful field trial but has not yet reached the market. The effectiveness of pulsed light treatments for surface microbial reduction is complicated by surface topography.

7.7.4.3. UV-C Radiation

Ultraviolet (UV) radiation lies in the wavelength range of 200-315 nm, The microcidal effect of UV light lies between 250 and 280 nm (the so-called UV-C range), with the optimal effect at 253.7 nm. At 253.7 nm, UV-C is an effective sterilizing agent for bacteria, yeast, molds and viruses. Both vegetative cells and spores are susceptible to UV-C radiation. The microcidal effect is due to the disruption of microbial DNA. Commercially, UV-C irradiation is often exclusively applied in conjunction with hydrogen peroxide. Recently, rare (noble) gas halides like krypton fluoride have been used in the production of UV (Excimer) lasers that operate at 248 nm. It is difficult to illuminate the interior of a carton with laser light because the carton must be manually moved into the laser beam in order to project the beam onto various interior sections. The sterilizing effect of UV rays are less as compared to H_2O_2.

7.7.4.4. Plasma

Nonthermal plasma (NTP) is electrically charged matter made up of highly reactive species such as gas molecules, charged particles in the form of positive and negative ions, free radicals, electrons and electromagnetic radiation (photons) at temperatures close to absolute zero. NTP and a low-pressure jet can be used to clean the packing materials' surfaces. A low-pressure microwave plasma reactor has been tested for the sterilization of PET bottles and NTP can be used to clean the surfaces of packaging materials. The plasma produced by the ignition of a combination of N_2, O_2 and H_2 gases inside a bottle contains both UV photons and transient antibacterial free radicals. A decrease of 10^5 CFU was achieved at a treatment time of 5 s. This cold sterilization method is currently being commercialized.

7.8. Packaging Material for Aseptic Packaging

Aseptic packaging materials require to fulfill a number of criteria for commercial uses:

- Packaging material should be suitable for sterilization.
- Must be inert and non reactive.
- Good barrier to the oxygen, light and water vapor.
- Good tensile strength.
- It should be of cost effective and easy to handle.

Table 1: Properties of packaging material used for aseptic packaging.

Material	Barrier property			Seal strength	Durability		
	Oxygen	*Moisture*	*Light*		*Stiffness*	*Tear*	*Puncture*
LLDPE		√		√		√	√
LDPE		√		√			
HDPE		√	√	√	√	√	√
PS					√		
PP		√				√	
PVdC	√	√					
EVOH	√						
PA (Nylon)						√	√
EAA				√			√
Paperboard			√		√		
Aluminum foil	√	√	√				
Metalized film	√		√				

LLDPE-Linear-low density polyethylene; LDPE-Low-density polyethylene; HDPE-High-density polyethylene; PS-Polystyrene; PP-Polypropylene; PVdC-Polyvinylidene chloride; EVOH-Ethylene vinyl alcohol; PA-Polyamide (Nylon); EAA- Ethylenacrylic acid

As indicated in Table 1, HDPE has highest barrier property and durability but in case of oxygen barrier property polyvinylidene chloride, ethylene vinyl alcohol, aluminum foil, metalized film are preferred. LLDPE has low stiffness but has similar properties compared to HDPE.

7.8.1. Classification of Aseptic Packaging Material

The packaging materials used for aseptic packaging are generally classified into rigid, semi-rigid and flexible packaging.

(i) ***Rigid packaging material*** are those whose shapes of the contours of filled and sealed container are neither affected by materials inside the packet and nor by the external mechanical pressure up to 70 kPa. Example of rigid packaging includes metal cans, glass container, and rigid cups.

(ii) ***Semi rigid packaging material*** are those whose shapes of the contours of filled and sealed container does not get affected by materials inside the packet under normal atmospheric pressure but can be affected by an external mechanical pressure of less than 70 kPa. Example of semi rigid packaging material includes those which are made of paper lamination, plastic films aluminum foil which are used by tetra pack etc.

(iii) ***Flexible packaging material*** are those whose shapes of the contours of filled and sealed container may affected by the materials inside the packet. Examples of such packaging materials includes bags and pouches which may or may not have exterior supports.

7.9. Aseptic Packaging System

There are numerous aseptic filling system available in market. These are designed to meet the requirements of aseptic packaging. These aseptic filling system differ on the basis of type of container and whether they are preformed or formed during filling. The basic properties of important aseptic filling system are mentioned in Table 2.

Table 2: Properties of aseptic package systems

Aseptic Packaging System	**Properties**
Cartons	• Cartoon material is laminate of paperboard coated inside and outside with polyethylene and an aluminum foil layer in between. • Available in form-fill-seal (formed from continuous reel within filler area) and preformed (Preformed, folded flat and assembled into cartoon in filler area) form. • Cartoon is sterilized by hydrogen peroxide and heat. • Sterilized cartoon are protected from entry of microorganism during filling and sealing by keeping in sterile zone maintained at an overpressure of 0.5 bar with sterile air. • Air used in aseptic zone is heated to 380 °C for sterilization and then cooled to 55 °C • Used for milk, cream, soybased milks, fruit juices, and recently for soups and sauces
Cans	• Made up of tinplate or drawn aluminium. Electrolytically chromium-coated steel and aluminum are now common. • Composite cans are made from aluminium foil, plastic and paper with metal ends. • The solder in tinplate cans may have to be of higher melting point than normal to withstand the can sterilization temperatures. • Are expensive, bulky to ship and store • Superheated steam at 200-220 °C is used sterilize the cans at atmospheric pressure in about 40 s. • Filled using a simple in-line filler of the slit or multiport type. • Available in volumes from 125 ml to 22 l
Bottles	• Made of glass, high-density polyethylene and polypropylene (PP) • Can be sterilized with all methods • Glass has weak or no light protection • Glass material can be mixed with pigments for light protection • Plastic bottles are available as nonsterile, and single station systems (blowing, filling, and sealing in one step) • H_2O_2 or peracetic acid (PAA) for sterilization • Blow molded plastic bottles are immune to thermal shock

Aseptic Packaging System	Properties
Sachets and pouches	• Made of laminate of low-density polyethylene, a center layer of ethylene vinyl alcohol (EVOH) and carbon black. • Heat-sealable plastic film is used, fed continuously from a roll. • Available in preformed and lay-flat tubing form. • Process of forming is similar to that used to form cartons from reels. • Film is sterilized in H_2O_2 bath. For 1-L pouch 20 s immersion in bath is required.
Cups	• Made up of high impact polystyrene, PP, polyvinylidene chloride, and EVOH • Available in preformed and form-fill-seal form. • Sterilization by H_2O_2 or a combination of both.

7.9.1. Carton System

The carton material is made of layers of unbleached and/or bleached paperboard that have both their internal and external surfaces coated with polyethylene. This creates a carton that is impermeable to liquids and allows for heat sealing of both the internal and exterior surfaces. Additionally, there is a 6.3 m-thick layer of aluminum foil that serves as a light and oxygen barrier. Figure 2 depicts the structure of a typical paperboard container. The following are the roles played by the various layers:

(i) The outer polyethylene (12 gsm) covers the ink layer and makes it possible to seal the package flaps.

(ii) The bleached paperboard (186 gsm), which also provides the package with the necessary mechanical strength, acts as a carrier for the decoration.

(iii) The polyethylene (20 gsm) laminating layer secures the aluminum to the paperboard.

(iv) The 6.3 mm thick (17 gsm) aluminum foil serves as a gas barrier and shields the product from light.

(v) The poly(ethylene-co-methacrylic acid) adhesive layer (6 gsm) ensures strong adherence between the polyethylene and the aluminum foil.

(vi) A heat-sealable liquid barrier is provided by the inner polyethylene layer (29 gsm)

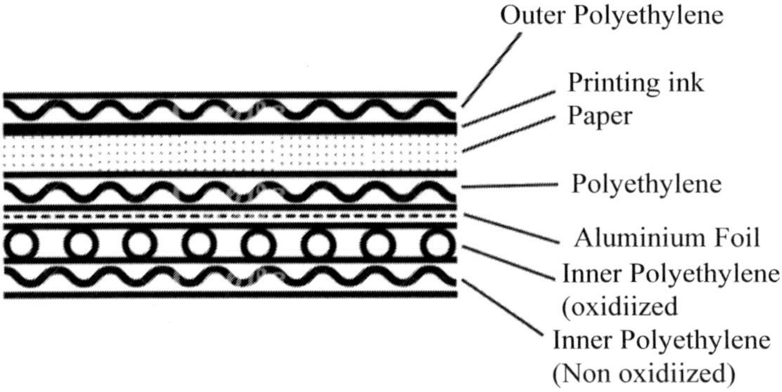

Fig. 2: Structure of paperboard laminate cartoon for aseptic packaging

-The "**Aseptic Bag-In-Box**" system is designed to package both "High" and "Low" acid items as well as products containing particles. Available fill volumes range from 25 liters to 1140 liters.

Aseptic bulk bags-A bulk aseptic bag is a multi-layered construction made up of an inner bag in contact with the product and an exterior barrier laminate. Gamma irradiation is used to pre-sterilize each bag, which is supplied flat.

7.10. Filling System

The most complex steps in aseptic processing is the filling of sterilized packaged or container with sterilized products. Before the process of filling start, the entire section is degassed to avoid any unwanted oxidative reaction. The process of filling involves exposure of food products to the vacuum in a tube and later storing the particular food products in a cooled tank. This tank system allows both sterilization and filling processes to be carried out independently of each other. For the purpose of filling, the product is initiated into the filling machine by sterile air, which expels the product out of the tank. This way product is filled in completely safe and sterile environment. The whole filling process is carried out in a hermetically sealed sterile environment. This is required to avoid product contact with the surrounding air during the filling process in the package. If no sterile air is used, microbes will be 'filled' in the package, thereby destroying the achieved sterilization.

7.11. Sealing of Container

The last but the most important steps in aseptic packaging is the sealing of container which must be hermetically sealed to ensure the sterility of packed food products over the time. Heat sealing is mostly done at defined temperature and pressure to melt down the sealing material present inside the packet and fill the entire cavities. Polyethylene (PE) is the most common sealing material which is found in both outer and inner layer of packaging material. PE is melted with the help of either sterile air or electric impulse.

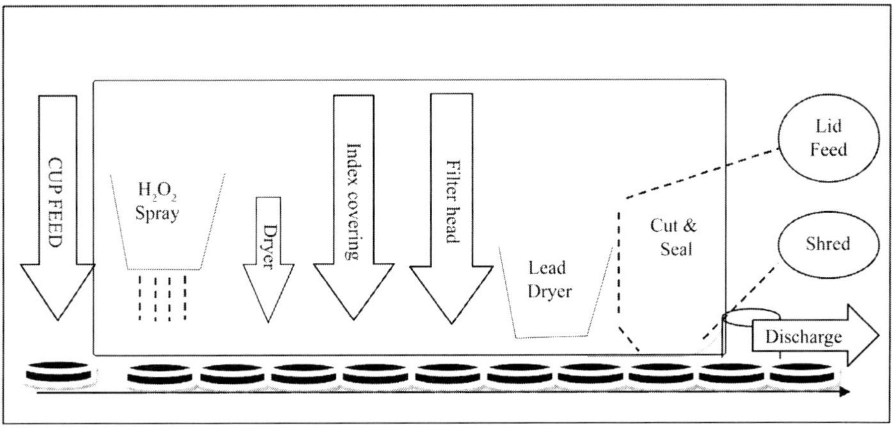

Fig. 3: Sterilization of packaging material with chemical followed by sealing and discharge

7.12. Safety Aspects of Aseptic Packaging

The process of aseptic packaging is prone to a series of contamination risks which includes risk associated with prepackaging, packaging and post packaging, for example:

(i) ***Post Process Contamination:*** Post-process contamination is the contamination of a product or material, typically by microbes, after it has undergone the finishing process. Post-process contamination is the most common form of microbial spoilage in pharmaceuticals and foods.

(ii) ***Insterile Air Flow:*** Air flow for aseptic packaging needs to be sterile in order to sustain the sterile atmosphere, Good quality filters or HEPA (*High efficiency particulate air*) filters are used in industries for the same.

(iii) ***Incoming Water Flow:*** Water is an essential raw material and can easily contaminate the product during preparation or usage/cleaning if not filtered properly. Filters/Regular sampling at various water points to ascertain the quality of water is important at frequent intervals.

(iv) ***Incomplete Product Sterilization:*** Temperature at the sterilization needs to be monitored regularly with regular calibration of temperature sensors.

(v) ***Pack Seal failure:*** Sealing heaters are to be regularly checked for failures.

(vi) ***Insufficient sterilization of packaging material:*** As with other raw materials, packaging material also needs to be aseptically handled, sterilized before coming in direct contact with the food products.

(vii) ***Unhygienic Conditions:*** Basic hygiene (GMP) needs to be maintained to ensure working of all the control measures. Failure of any basic hygienic condition can lead to failure of all control measures. Some of the important risk are being schematically represented in Figure 4.

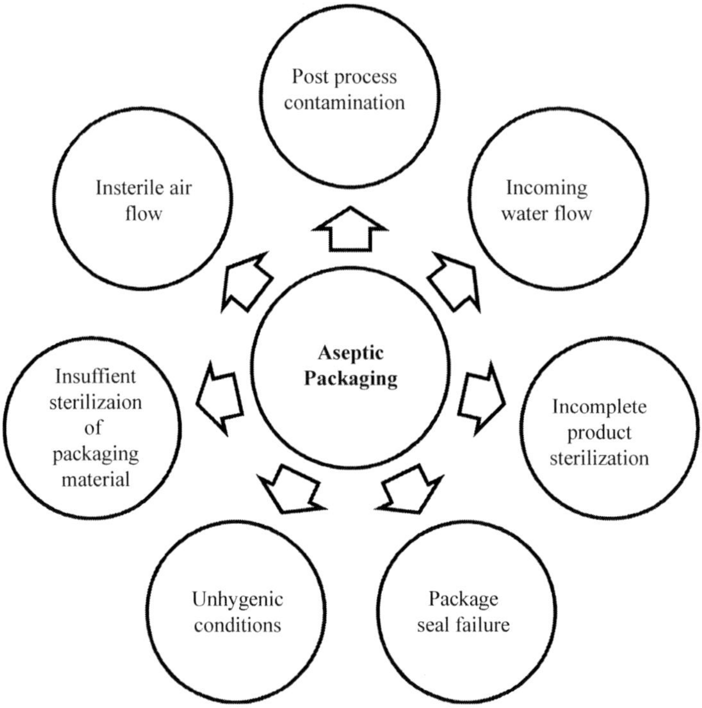

Fig. 4: General risk affecting aseptic packaging processing. Adopted from Encyclopedia of Food Safety, Vol: 3, Page: 124-134

7.13. Advantages of Aseptic Packaging

(i) Aseptically packaged food retains quality, nutrients and sensory attributes. The retention of vitamins is particularly high. Bioactive components and natural pigments such as anthocyanin and chlorophyll are better retained in HTST process used in aseptic processing.

(ii) Aseptic packaging improves the mobility and transportation of foods.

(iii) Aseptically processed and packaged products can be stored at room temperature.

(iv) It is possible to fill sterilized products into containers of different types and shapes attractive to consumers.

(v) Aseptic packaging can yield microwaveable products which is a consumer friendly feature. This can command two to three times the price of that of similar product in metal cans

(vi) There is energy saving of 30% or more in aseptic processed and packaged food as compared to retort processing.

References

Ansari, I. A., and Datta, A. K. (2003). An overview of sterilization methods for packaging materials used in aseptic packaging systems. Food and Bioproducts Processing, 81(1), 57-65.

Bertolini, M., & Romagnoli, G. (2012). An Italian case study for the Process-Target-Cost evaluation of the ohmic treatment and aseptic packaging of a vegetable soup (minestrone). Journal of food engineering, 110(2), 214-219.

Chuaqui-Offermanns, N. (1989). Food packaging materials and radiation processing of food: a brief review. International Journal of Radiation Applications and Instrumentation. Part C. Radiation Physics and Chemistry, 34(6), 1005-1007.

David, J. R., Coronel, P. M., Simunovic, J., Graves, R. H., Carlson, V. B., & Szemplenski, T. E. (2022). Aseptic Processing and Packaging. Handbook of Aseptic Processing and Packaging, 1.

Fernández, A., Soriano, E., López-Carballo, G., Picouet, P., Lloret, E., Gavara, R., & Hernández-Muñoz, P. (2009). Preservation of aseptic conditions in absorbent pads by using silver nanotechnology. Food research international, 42(8), 1105-1112.

Floros, J. D. (1993). Aseptic packaging technology. W: Principles of aseptic processing and packaging. Red. PE Nelson, JV Chambers. The Food Processors Institute. Washington, DC.

Götz, A., Wani, A. A., Langowski, H. C., & Wunderlich, J. (2014). Food technologies: Aseptic packaging.

Gordon L. Robertson, (2013) Food Packaging Principles and Practice. CRC Press.

Singh, R. K., & Singh, N. (2005). Quality of packaged foods. In Innovations in food packaging (pp. 24-44). Academic Press.

Moruzzi, G., Garthright, W. E., & Floros, J. D. (2000). Aseptic packaging machine pre-sterilisation and package sterilisation: statistical aspects of microbiological validation. Food Control, 11(1), 57-66.

Mostert, M. A., Buteux, G., Harvey, P. C., Hugelshofer, W., Mellbin, P., Nassauer, J. & Wilke, B. (1993). Microbiologically safe aseptic packing of food products. Trends in Food Science & Technology, 4(1), 21-25.

Pruß, K., Stirtzel, S., &Kulozik, U. (2012). Influence of the surface temperature of packaging specimens on the inactivation of Bacillus spores by means of gaseous H2O2. Journal of applied microbiology, 112(3), 493-501.

Ros-Chumillas, M., Belissario, Y., Iguaz, A., & López, A. (2007). Quality and shelf life of orange juice aseptically packaged in PET bottles. Journal of Food Engineering, 79(1), 234-242.

Sandeep, K. P., Zuritz, C. A., & Puri, V. M. (1999). Determination of lethality during aseptic processing of particulate foods. Food and Bioproducts Processing, 77(1), 11-17.

8

Active and Intelligent Packaging

Rahul Das[1], Yogesh Kumar[1] and Vijay Singh Sharanagat[2]

[1]Department of Food Engineering and Technology, Sant Longowal Institute of Engineering & Technology, Longowal, Punjab
[2]Department of Food Engineering, National Institute of Food Technology Entrepreneurship and Management, Sonipat, Haryana

8.1. Introduction

Food packaging's essential functions is to protect the food from hazardous compounds, and contaminant. In order to improve the shelf life of food commodity, smart packaging like active and intelligent packaging has been developed which can monitor freshness, display information of quality, and improve food safety. More precisely, smart packaging was defined by *Smart Packaging Journal* as "*the use of features of high added value that enhance the functionality of a product, notably mechanical, electronic and chemical features that improve safety and efficiency*". The major concerns with traditional packaging are not only shelf life of packaged food, but also the inability of packaging material to convey the important information about quality changes to both customer and manufacturer. On the other hand, smart packaging enables the manufacturer to monitor and control the quality changes of food in the supply chain, and reinforce safe handling practices. Smart packaging techniques are classified into three categories based on the function as shown in the flow diagram (Figure 1).

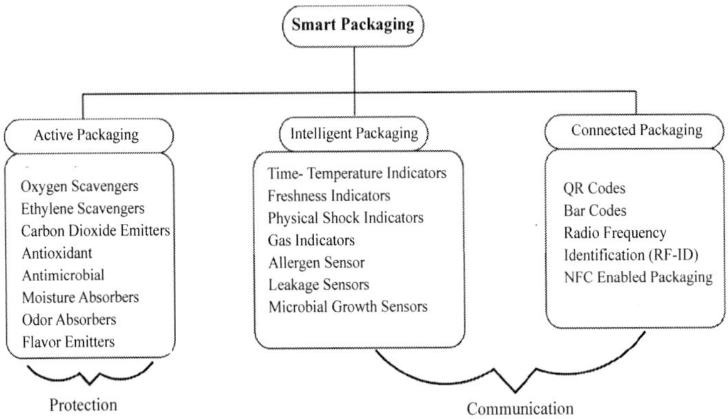

Fig. 1: Classification of smart packaging techniques for food-based products.

All three types of smart packaging techniques have different functions; however, it is very common to see all three at the same time. Therefore, by using active, intelligent, and connected packaging, both manufacture as well as customer can get important information about the package. Some of advantages of active and intelligent packaging is given in Table 1.

Table 1: Advantages of active and intelligent packaging

Active packaging	Intelligent packaging
• Ability to prevent undesired changes like oxidation phenomena (such as rancidification of fats and oil), consequent off-odors and off flavors, color characteristics as well as loss of oxygen sensitive nutrients. • Can inactivate or destroy the growth of aerobic microorganism into the food product. • Eliminates/minimizes the need for preservatives and antioxidants. • Reduces the metabolism rate in the food. • Significantly increases the shelf life by fixing various chemical changes in packaged food. • Minimizes the loss and waste of food by reducing nutritional degradation in harsh environmental conditions like desert.	• Capable of carrying out the intelligent functions such as detecting, sensing, tracking, recording, communicating, and applying scientific logic. • Alerting manufacturer and consumer by indicating possible problems. • Provide the information about the integrity of the package as well as condition of the food product throughout supply chain. • Smart monitoring using intelligent packaging helps in reducing food loss and waste and enhance bio-security. • Enables manufacturer to optimize the different factors from manufacturing to consumption.

8.2. Active Packaging

Active packaging is an excellent example of innovative technology that exceeds the traditional function of packaging food to enhance the product shelf life or improve its safety and sensory properties by sensing and modifying the changes in the internal and external environment. According to European regulation (EC) No 450/2009, active packaging systems are designed to "*deliberately incorporate components that would release or absorb substances into or from the packaged food or the environment surrounding the food.*" For example, unwanted compounds such as ethylene (C_2H_4) produced by respiring fruits or O_2 present inside a package must be removed with the help of active agents to increase the shelf life and maintain the overall quality. The removal of such compounds is possible because active materials inside or within packaging materials react with these undesired compounds. On the other hand, CO_2 or ethanol are used to inhibit microbial growth, change the film permeability, and modify the package's physical conditions. Golden L. Robertson, the author of the most popular and well-known book "Food Packaging Principles and Practice,"defined active packaging as "*packaging in which subsidiary constituents have been deliberately included in or on either the packaging material or the package headspace to enhance the performance of the package system*". Therefore, a packing will be never considered as active unless the packaging actively affects its internal atmosphere of package. For example, MAP (Modified atmospheric packaging) is a normally practiced passive packaging in which inert gas mixture is delivered immediately before sealing and the air is removed from the package. In case of MAP, the role of inert gases is not to actively modified the internal atmosphere, but to replace excess oxygen to restrict oxidative degradation of food inside.

8.3. Classification of Active Packaging

Active packaging can be classified into two major systems: active scavenging systems (absorbers) and active releasing systems (emitters). The example of active scavenging system includes the removal of moisture, CO_2, O_2, ethylene, odor, and other unwanted compounds. The examples of active releasing system are carbon dioxide, antimicrobial compounds, antioxidants, flavors, ethylene, or ethanol emitters to the packaged food. Some common active packaging systems and their properties and mode of action in the food packages are given in Table 2, Table 3 and Table 4.

Table 2: Different properties of active package.

Properties	Examples
Absorbing/ scavenging properties	Oxygen, carbon dioxide, moisture, ethylene, flavors or odor, taints, UV light
Releasing/ emitting properties	Ethanol, carbon dioxide, antioxidants, antimicrobial, preservatives, sulfur dioxide, flavors
Removing properties	Catalyzing food component removal: lactose, cholesterol
Temperature control	Insulating materials, self-heating and self-cooling packaging, temperature-sensitive packaging
Microbial and quality control	UV and surface-treated packaging materials

Table 3: Types of active food packages, their effects, and function on the packaged food.

Types of Food Package	**Function**	**Results/ effects**
Oxygen Scavenger	Minimize the effects of oxygen permeation through the packaging materials during product shelf life, preserve the quality of food product, inhibition of undesirable oxidation (unstable pigments), aerobic microorganism growth, prevent oxidative rancidity, controlling enzymatic discoloration, inhibition of oxidation of lipids, development of molds, preventing discoloration	Extends the shelf life of Cheese Spread, increase the shelf life in cakes, quality improvement and increase the shelf life in dark chocolates, inhibition of ascorbic acid degradation in beverages, antimicrobial activity against rhizopus stolonifer in Sliced bread.
Antimicrobial	Prevent aerobic microorganism growth, prevent oxidative rancidity, control enzymatic discoloration, inhibit oxidation of lipids, development of molds, prevent discoloration	Antimicrobial activity against *Lactobacillus acidophilus*, *Pseudomonas fluorescens, Listeria innocua,* and *Escherichia coli* in raw fish products and salmon, antimicrobial activity against filamentous fungi and yeast in butter, increase the shelf life in fresh beef, increase the shelf life in grilled pork.
Antioxidant	Inhibition of oxidation processes	Improvement of quality in milk powder, quality improvement of apple slices
Odor releaser and antimicrobial	Odor stabilization	Quality improvement and antimicrobial activity in roast beef
Ethylene scavenger	Reduction in ripening and senescence, deterioration, extend shelf life, control of ripening of fruits and vegetables	Reducing the rate of softening in tomatoes
Moisture scavenger	Extension of shelf life through maintaining moisture content, decrease in moisture condensation in the packaging, positive impact on the appearance, reduction in browning or discoloration	

Table 4: List of common active packaging systems with examples and their applications in food industries.

Active packaging system/ action	Mechanisms/ substances	Food application
Oxygen absorbers or scavengers	Iron-based, metal/acid, metal (e.g., platinum) catalyst, ascorbate/metallic salts, enzyme-based, and nylon MXD6	Bread, cakes, cooked rice, biscuits, pizza, pasta, cheese, cured meats and fish, coffee, snack foods, dried foods, and beverages
Ethylene absorbers	Potassium permanganate, activated carbon, and activated clays/zeolites	Fruits and vegetables
Moisture absorbers	Polyvinyl alcohol, encapsulation, silica gel, poly (vinyl acetate) blanket, activated clays, and minerals	Fish, meats, poultry, snack foods, cereals, all dried foods, sandwiches, fruits, and vegetables
CO_2 absorbers/ emitters	Iron oxide/calcium hydroxide, ferrous carbonate/metal halide, calcium oxide/ activated charcoal and ascorbate/sodium bicarbonate	Coffee, fresh meats and fish, nuts, other snack foods and sponge cakes, and fruits and vegetables
Flavor/ odor absorbers	Cellulose triacetate, acetylated paper, citric acid, ferrous salt/ ascorbate, sodium bicarbonate, and activated carbon/clays/ zeolites	Beverage product (fruit juices), prepared food like fried snack foods, fish, dry cereals, poultry, dairy products and fruits, frozen food like ice cream
Ethanol emitters	Encapsulated ethanol, micro-encapsulated ethanol	Prepared foods such as pizza crusts, cakes, bread, biscuits, bakery products, semi-dry fish, meat, and fruits, dairy product (cheese), and sweet baked goods
Antimicrobial releasers	Sorbates, benzoates, propionates, sliver salts, sulfur, and mercurial compounds, bacteriocins, sub-micrometer cell wall penetrates, and zeolites	Minimally processed fruits, cheese, meat, bakery products such as bread, cakes
Antioxidant releasers	BHA/ BHT, TBHQ, and vitamin C or E	Breakfast cereals, bakery products (Hard baked goods), beverages (Bag-in-box wine, wine)
AM packaging	Organic acids, silver zeolite, spice and herb extracts, BHA/ BHT antioxidants, vitamin E antioxidant, chlorine dioxide, and sulfur dioxide	Cereals, meats, fish, bread, cheese, snack foods, fruits, and vegetables
Self-heating and self-cooling	Quicklime/water, ammonium nitrate/water, and calcium chloride/ water	Ready meals and beverages
Changing gas permeability	Side-chain crystallizable polymers	Fruits and vegetables

So far, we have seen various properties, applications and effect of active agents, but how to use them is also important as most of them are chemical active substances. If they come in direct contact with foods, food may become harmful for consumption. To overcome this problem, active ingredients are either incorporated directly into the packaging film and or placed inside the sachets/pads and attached to the inside wall of the package. Sachets and pads are extensively used for solid foods as there is very less chance of active substance kept in sachet coming in contact with food. Moreover, if they contain substances which are not harmful for human consumption, they can also be used in liquid foods. However, they are generally avoided in liquid foods as fluid can penetrate inside them and react to form undesired substances. Hence, active substances are generally placed in the headspace of container to avoid any direct contact. Despite few drawbacks, the absorbers or releasers are widely used in active packaging, and their important functions or mode of action are discussed below.

8.3.1. O_2 scavengers

In most cases, the deterioration and spoilage of food are mainly due to the oxidation of food constituents. Therefore, most of the oxygen-sensitive food are packed using modified atmospheric packaging (MAP) or vacuum packaging. However, it is very hard to remove oxygen from packaged food completely. Therefore, an oxygen scavenger is used for O_2 sensitive food to absorb the residual part of O_2 after packaging to improve the quality, safety, and shelf life. In general, O_2 scavenging technologies use one or more of the following concepts: iron powder oxidation, ascorbic acid oxidation, and enzymatic oxidation. The working principle of iron oxidation based O_2 scavengers is shown below:

$$Fe \rightarrow Fe^{2+} + 2e^{-}$$

$$\frac{1}{2}O_2 + H_2O + 2e^{-1} \rightarrow 2OH^{-1}$$

$$Fe^{2+} + 2\ OH^{-1} \rightarrow Fe\ (OH)^2$$

$$Fe(OH)_2 + \frac{1}{4}O_2 + \frac{1}{2}H_2O \rightarrow Fe(OH)_3$$

Ascorbic acid O_2 scavenging component is incorporated with the barrier packaging systems such as plastic or metal closures, crown caps and the ascorbate is oxidized to form dehydroascorbic acid and sulphite to sulphate. So, the crown caps are used in beer to prevent oxidation of flavors. Similarly, glucose oxidase or oxidoreductase is an oxygen scavenging enzyme that transfers of two hydrogen molecules of the -CHOH group glucose to O_2 and form glucono-δ-lactone and H_2O_2. Lactone simultaneously reacts with H_2O to form gluconic acid. Catalase is then introduced to break down the hydrogen peroxide to form H_2O and O_2.

The reactions are:

$2G + 2O_2 + 2H_2O \rightarrow 2GO + 2H_2O_2$ Where, G is the substrate.

$2H_2O_2 + \text{catalase} \rightarrow 2H_2O + O_2$

8.3.2. Carbon Dioxide Generating System/ Absorbing System

The generation of high CO_2 levels (60 to 80%) in a packaged food inhibits the growth of microorganisms on surfaces that prolong the shelf life of packaged food. However, the high level of CO_2 can cause a considerable change in the taste of food products. Therefore, CO_2 generators are used only in specific food products such as meat, fish, poultry, and cheese packaging. The emitters usually come in the form of pad or sachets. The absorbent pad (based on either ascorbic acid and ferrous carbonate or ascorbic acid with sodium bicarbonate) absorbs oxygen coming out of the product which reacts with the active ingredients inside the pad, resulting in the release of CO_2. This system has continuously produced CO_2 to maintain the desired concentration within the package, and in most of the plastic films, its permeability is 3 to 5 times higher than that of O_2. Therefore, selecting a better impermeable plastic film is suggested to reduce the loss of CO_2 and maintain the desired level.

On the other hand, most foods release CO_2, and after some period, it deteriorates. So, CO_2 scavengers such as physical absorbent (zeolite or an active carbon powder) and chemical absorbents (Ca $(OH)_2$, and Mg $(OH)_2$) are extensively used in small pouches of perforated polypropylene coated paper. Physical absorbents contain microporous structure that have some degree of CO_2 absorption capacity but they don't react with CO_2. The surface area of the pore, pore volume, pore size distribution are the important properties for absorption of gas. For chemical absorbent, the addition of calcium hydroxide in a sachet absorbs the CO_2 released from the food and improves the food quality and prolongs shelf life.

$Ca(OH)_2 + CO_2 \rightarrow CaCO_3 + H_2O$

8.3.3. Ethylene Scavenger

Ethylene is a phytohormone that initiates and enhances ripening, softening of structure and degradation of chlorophyll, thus deteriorating the overall quality and reducing the shelf life of freshly harvested fruits and vegetables. During the storage of fruits and vegetables, controlling ethylene is very important to prevent them from ripening after harvest. Therefore, ethylene scavengers are very useful in ethylene sensitive fruits and vegetables such as apples, bananas, mangos, tomatoes, onions, carrots.

Potassium permanganate is a common ethylene scavenger, used to oxidize the ethylene to carbon dioxide and water and restrict the ripening. During this process, potassium permanganate changes color from purple to brown, indicating the absorbing capacity of the remaining part of ethylene. But, potassium permanganate ($KMnO_4$) is a toxic compound, so its direct use should be avoided in cases of food.

$$3C_2H_4 + 12KMnO_4 \rightarrow 12MnO_2 + 12KOH + 6CO_2$$

8.3.4. Flavor and Odor Absorber/ Releaser

Addition of essences and odors in a food can improve the flavor and aroma of the fresh product as well as increase the demand of the food to the consumer. The release of odors from a food package indicates the loss of smell and taste. An active packaging technology to reduce the bitterness in citrus fruit juices has been developed. An active thin cellulose acetate (CA) layer is applied inside the packaging materials which contains the fungal-derived enzyme naringinase, consisting of α-rhamnosidase and β-glucosidase. This enzyme hydrolyses naringin (it is found in most citrus food causes bitter taste) to naringenin and prunin, both are non-bitter compounds. Flavor-enriched packaging materials are also used to emit the desirable flavor into the food that improve the organoleptic product quality. Two types of taints (amines and aldehydes) could also be removed by active packaging. Malodorous amines cause breakdown of protein in fish muscle. The ANICO bags (ANICO CO., Japan) are the films consisting of ferrous salt and organic acid such as ascorbic or citric acid that are used to oxidize the amines and other oxidizable odor-causing compounds as they are absorbed by the polymer.

8.3.5. Antimicrobial Preservative Releasers

Antimicrobial food packaging materials are used to reduce the microbial growth which in turn enhances the shelf life and maintain the quality as well as safety of the product. The basic principle of antimicrobial films is based on the release of antimicrobial entities. The major food application of antimicrobial films includes fish, meat, poultry, cheese, bread, fruits, and vegetables. The most common antimicrobial agent used in Japan is Ag-substitute zeolite, which is incorporated into plastics. The usage of Ag-ions (strong antimicrobial activity) inhibits a range of metabolic enzymes. Other chemical preservative that have been introduced in active antimicrobials releasing systems including organic acid such as propionate, sorbate, and benzoate, bacteriocins (such as nisin and pediocin), enzymes (such as lysozyme, glucose oxidase, lactoperoxidase), metals, fungicides (imazalil and benomyl), extracts from spices and herbs, SO_2 and chlorine dioxide. A typical four layer packaging films consisting of outer layer, barrier layer, matrix layer, and control layer is used in the food. The matrix layer contains antimicrobial substance and its release to the food surface is controlled by the control layer. The working principle of this multi-layer packaging film is shown in figure 2.

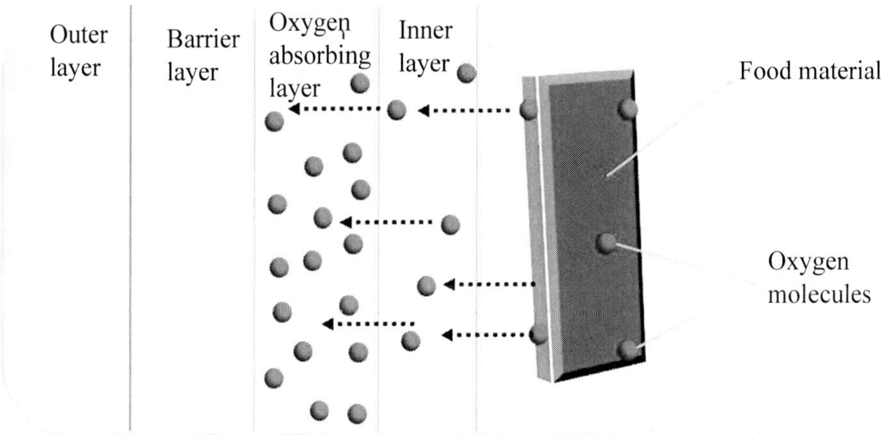

Fig. 2: The working of atypical multi-layer oxygen absorbing active packaging film.

8.3.6. Moisture / Relative Humidity Controller Films/Sachets

In most of the cases, some of the moisture remains inside the package during packaging of foods which may be absorbed by the product or form condensate causing microbial spoilage or poor consumer perception. This problem can be minimized by using moisture or relative humidity controller films or sachets. Moisture controller film ensures controlled permeability of water vapour to maintain a fixed relative humidity inside package. Moisture controlling sachet can absorb the moisture in the package. For example shelf life of packaged tomatoes was extended from 5 to 15-17 days by using NaCl sachet which lowered the water activity inside the package and preventing mold growth. Another approach to reduce the moisture inside package is to intercept the water vapour inside package by placing humectants between two layers of plastic film. These humectants absorbs the remaining moisture inside the sachets in order to enhance the shelf life of the food product. An example is PichitTM film (Showa Denko, Japan) used for wrapping of fish and chicken that controls the equilibrium relative humidity (ERH) of the product. Several materials are often provided in TyvekTM sachets such as silica gel, calcium oxide (CaO) and montmorillonite.

8.3.7. UV-light Absorbers

The packaging materials are used to protect the ultraviolet rays (causes photo oxidation reaction) in order to avoid the food (such as dairy products, nuts, meat products, and wine) from deterioration. The light absorbers or blocking materials such as organic (Phenolic-type UV absorbers), inorganic (TiO_2, CeO_2, Fe_2O_3, ZrO_2 and ZnO), hybrid organic-inorganic (Phenolic-type UV absorbers are blended with silicon alkoxides), and natural light absorbers have been embedded into the plastics (Figure 3). One of the most important UV-light absorbers is organic compounds. The phenolic-type UV absorbers consist of hydrogen bonds from either by O-H-O bridges or by O-H-N bridges. The intramolecular H-bonds convert the incident light energy to low thermal energy by means of photo physical reactions which decrease the free radical formation in the early stage of degradation.

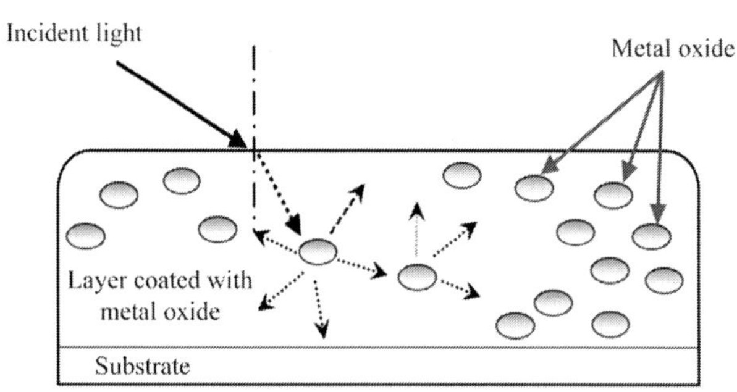

Fig. 3: Schematic of absorption and scattering of light with metal oxide.

8.3.8. Lactose Remover

In today's society with growing trend of health awareness, it has become desirable to remove particular compounds from food (such as dairy products) including lactose for the benefit of lactose intolerant people. The most common beta-galactosidase or lactase enzyme can be applied in active packaging materials (surface functionalized low density polyethylene films). The lactase can be covalently attached with the wall of the packaging materials that has been designed to reduce the lactose content of milk and other dairy products by hydrolysis of lactose during storage.

8.3.9. Cholesterol Remover

In general, the cholesterol reductase enzyme is used in active packaging materials for liquid based products. This enzyme is used to convert cholesterol into coprosterol to make the product cholesterol free.

8.3.10. Antioxidant Releasers

Lipid oxidation is one of the main factor causing food spoilage, microbial growth, reduction in shelf life, change in taste or odor, deterioration of texture as well as reduction in nutritional quality of the food products. So, antioxidants are widely used in the food packaging systems and they act as a food additive to improve the oxidative stability. Butylated hydroxytoluene (BHT), butylated hydroxyanisole (BHA) are some example of synthetic antioxidant that can be incorporated into plastic films (such as LDPE, and HDPE/EVOH/LDPE). Natural antioxidants such as tocopherols, polyphenols, plant extracts, and essential oils are now growing interest among the researchers to be used in the food packaging systems. The antioxidant releasers are used in food such as oil, nuts, bakery products, fresh meat, fruits and vegetables to prevent the formation of radicals, mainly oxo, hydroxyl, and superoxide that inhibits the oxidation reaction and delay of protein denaturation.

8.3.11. Self-heating Aluminum or Steel Cans and Containers

Self-heating cans or containers are used as an active packaging material with the ability to heat food products without any external heat sources or power. Exothermic reaction is initiated inside the self-heated cans food products with the help of pressing on the bottom of the can by user. It is generally made as a triple-walled container and used in beverages. The working principle for generation of heat is that when the user applies the force (by pressing) on the bottom of the can, a rod pierces the membrane, allowing the heating agent and the water to well mix. The resulting mixture releases heat and thus warms the beverages. There are some examples of heating agents such as calcium oxide, copper sulphate, and anhydrous calcium chloride. The reaction of calcium oxide is given below:

$$CaO + H_2O \rightarrow Ca(OH)_2$$

8.3.12. Self-cooling Aluminum or Steel Cans and Containers

Self-cooling active packaging system is a new emerging technology, used to provide chilled products without cooled storage. The design of self-cooling beverages consist of a tin-plate steel outer can to contain the beverage and an inner aluminum can (known as Heat Exchange Unit (HEU)) that contain the absorbent. The advantage of this system is to reduce the emission of greenhouse gases. In beverages systems, the HEU contains the activated carbon with absorbed CO_2 that provides the cooling effect as well as prevents contact between the beverage and the activated carbon. The cooling effect is generated with the help of valve that releases pressure inside the HEU by venting the CO_2 to the atmosphere (desorption of CO_2 is endothermic). The temperature of beverage drops by 15 °C in only 90 seconds through this mechanism.

8.4. Intelligent Packaging

The development of intelligent packaging represents a packaging system that can detect environmental changes such as moisture and temperature changes inside the packaged food, and inform the same to the users. All the information of the product (origin, date, conditions of storage, growth of microorganism, theoretical expiration, the composition of the product, etc.) is provided into this packaging system. The product information is obtained from sensors, indicators, and or devices used in this packaging system. The different types of intelligent packaging system used are shown in Figure 4.

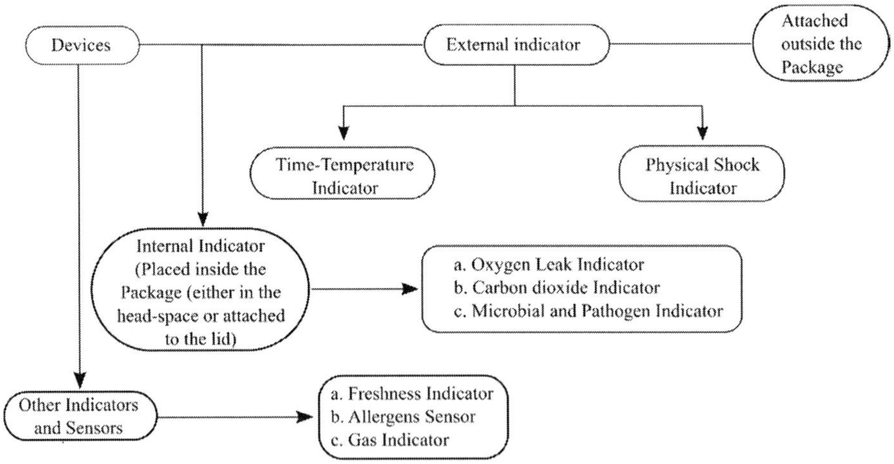

Fig. 4: Flow chart of different indicators used in food packaging.

Indicators are the devices used to determine the efficient flow of information and communication between the product and the consumer. The product has some specific bars which store the information about the product from use to consumption date.

8.4.1. Time-Temperature Indicators (TTIs)

Time-Temperature indicators (TTIs) are defined as an integrator designed to monitor the time and temperature (e.g., below or above a reference critical value) continuously throughout the food chains. The basic principle of the operating system is based on the chemical, electrochemical, mechanical, microbiological, and enzymatic change (oxidation of phenols, sugar-starch conversion) during distribution and storage in the form of deformation due to mechanical change or development and movement of color.

TTIs can be classified into two major categories, which are partial history indicators and full history indicators. Partial history indicators responds to when the predetermined level of threshold temperature is exceeded (Figure 5). Full history indicators monitors an integrated time-temperature combination of entire product life which shows all the temperature continuously.

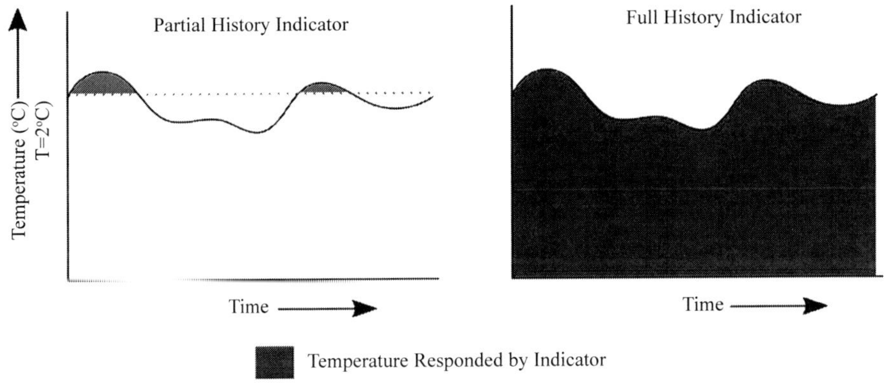

Fig. 5: Types of TTIs use in food packaging.

Any chemical reactions inside food are temperature-dependent phenomena and reaction will not start until reactants get the activation energy. Based on this fact, a temperature indicator utilizes chemical reaction to show color if it is exposed to higher temperature than designed temperature. The change in indicator color from its reference value is monitored with passage of time on a strip, as demonstrated in the Figure 6.

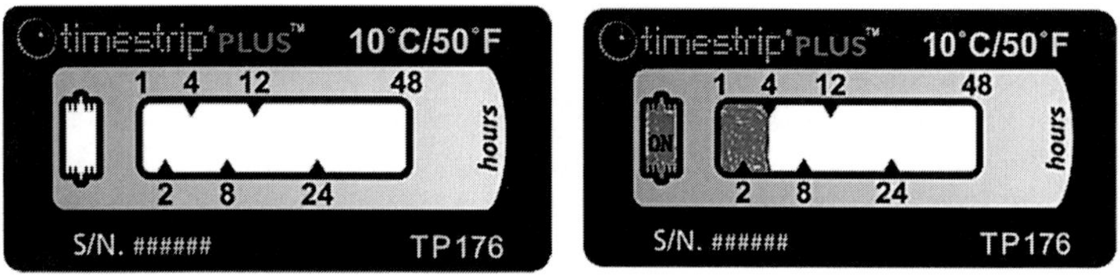

Initially — When exposed to temperature >10°C

Fig. 6: Development and movement of color when the indicator is exposed above a fix temperature. (Timestrip® plus is manufactured by Timestrip UK Ltd.)

8.4.2. Freshness Indicators

The freshness indicators are used in the packaged food to monitor the product's quality attributes or directly show the actual spoilage, growth of microorganisms, and metabolism of packaged goods. There are different varieties of freshness indicators that have been applied in the food packaging industry. The freshness indicators like bromothymol blue pH dyes have been frequently applied in the meat packaging industry to monitor the CO_2 formation into the packaged products. Other pH dyes are bromocresol purple, bromocresol green, xylenol blue, cresol red, phenol red, methyl red, and alizarin, which have been used for the same purpose. Freshness indicators can also be used to determine the remaining shelf life of perishable product as demonstrated in figure 7. These indicators placed inside sealed package changes color if a chemical change occurs within the product. They are sensitive to specific by-products that originate from deterioration reactions in the food. Initially, the main indicator remains light in color as of no reaction. Due to release of specific unwanted substances indicator starts becoming darker. We can use the product until the indicator color matches with references color. If the indictor color is darker, one should avoid the packaged food.

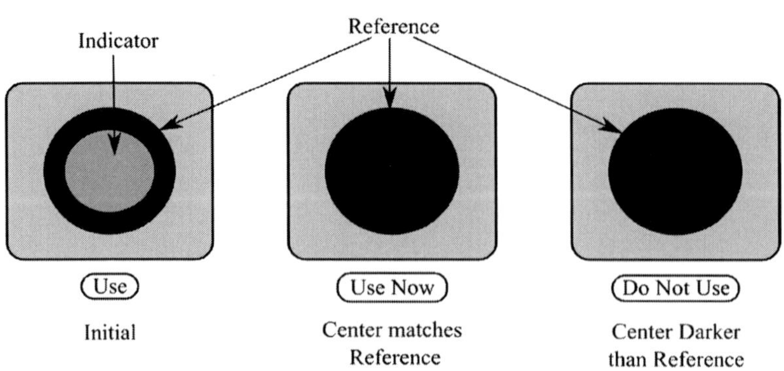

Fig. 7: Working principle of freshness indicator.

8.4.3. Physical Shock Indicators

It is very much possible for a food package to expose to high impact or shock during transportation, storage, and handling. These kinds of physical shock can significantly affect food quality, especially if food is liquid or semisolid. Physical shock indicators are effectively used to get a visual indication about whether package was exposed to a shock threshold or not. They change color after exposing to more impact than designed value, as shown in Figure 8.

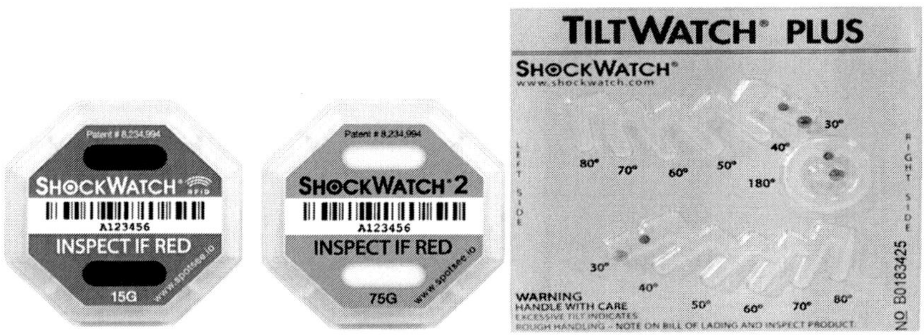

Fig. 8: Shock and tilt indicators are used to display the package is exposed to shock/ tilt or not. (ShockWatch, Inc. (Australia) is a leading manufacturer for various shock and tilt indicators.)

8.4.4. Gas indicators

Gas indicators are an alternative, non-invasive approach in intelligent packaging, which is extensively used in MAP (modified atmospheric packaging) to determine package integrity (e.g., seals leakage). These indicators respond to the changes of gas concentration or composition in a packaged food. Oxygen indicators are the most common gas indicators, used in the food packaging industry, where it detects the deterioration of food due to oxidative rancidity, change in color, and microbial spoilage. Oxygen indicators are used in the packaging films that shows the color change when changes in the gas composition due to enzymatic and chemical reactions (Figure 9). They also provide information about the concentration of altered gas such as O_2, CO_2, water vapor, ethanol, etc., in a qualitative as well as in a semi-quantitative way.

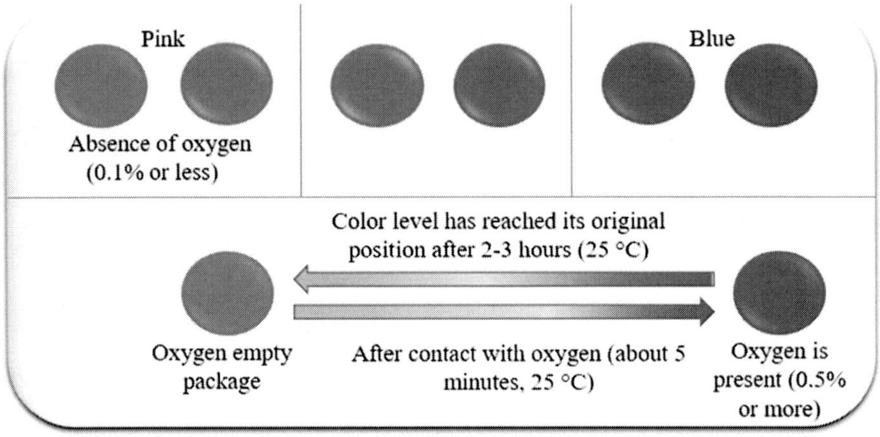

Fig. 9: Oxygen indicator tablet used for indicating either package was exposed to oxygen or not. (Redesigned from Ageless Eye® oxygen indicator which is developed by Mitsubishi Gas Chemical Company, Japan).

8.4.5. Leakage Sensors

The application of leakage sensors is to determine the food integrity throughout the food distribution chain. It is used in MAP for meat products to give information about O_2 and CO_2 leakage through the packaged product.

8.4.6. Microbial Growth Indicators

The microbial growth indicators are used to determine food product deterioration or microbial spoilage of food. These indicators are commonly used in perishable food such as meat, fish, and poultry products.

8.4.7. Ripeness Indicators

Ripeness indicators are introduced in the food packaging materials that gives us the information about the age of the food product.The example of this indicator is cheese ripening indicators that is used for spotting metabolites (majorly volatiles) or chemically breakdown the molecules by the process of glycolysis, proteolysis, and lipolysis to determine the maturity or age of any cheese verity.

8.4.8. Oxygen indicators

The oxygen or gas concentration indicators are placed inside the package to detect the change inside atmosphere. These changes occur due to penetration growth of microorganisms, enzymatic and chemical reactions in the food matrix. A typical oxygen indicator consists of a redox-dye (such as methylene blue), a reducing compound (such as reducing sugar), and analkaline compound (such as NaOH, KOH). O_2Sense™ is an eye readable indicator that shows a clear color change when the amount of oxygen presents inside the packaging films reaches to a predetermined value. This indicator is introduced in the packaging materials to alert producers, retailers, and consumers for the condition of the products.

8.4.9. Carbon Dioxide Indicator

Carbon dioxide indicators are used in the packaging materials to monitor and communicate the information about the quality of the product. It is categorized into two groups: (a) luminescent dyes-based indicators, when it is exposed to CO_2 then their color may change by changing its fluorescence property, (b) colorimetric indicators based on a pH indicator (pH dyes such as m-cresol purple) that have been used to generate visible color changes.

8.4.10. Pathogen Indicators

The pathogen indicators are used to monitor or detect the growth of microorganism (such as pathogenic microorganism that causes various diseases endangering human health) in products such as packaged meat, fish, or poultry. Toxin Guard™ is a commercially available systems to build polyethylene (PE)-based packaging materials, which can detect presence of pathogenic microorganism (such as *Salmonella* sp., *Escherichia coli* O157, *Campylobacter* sp., and *Listeria* sp.). Another example of this indicators is Food Sentinel System™, which is able to detect the specific microorganisms like *Salmonella* sp., *E coli*, and *Listeria* sp. A deep blue color polydiacetylene packaging material is used to bind or trap the toxin (originated from from) into the film matrix and therefore the color of the film changes from blue to red.

8.5. Role of RFID in Intelligent Packaging

Radio frequency identification tags (RFID) is a wireless (non-contact) data communication system, where tags are programmed with unique features and it is used for the purpose of identification and traceability of objects (products or packaging) along the supply chain. A typical supply chain includes the products from production, transportation, distribution, retail, and finally end to the consumer. So, from these points of view, the traceability of logistics flow is most important for the manufacturing process design, system layout, and improve security in work areas. In this system, a reader captures data from an RFID tags by emitting radio waves and the data is then passed onto a host computer (which may be connected to the internet or the local network) for making decision and analysis purpose (Figure 10). RFID tags are divided into two types: i) passive tags that have no battery, but the power is generated due to supply of energy by the reader, and ii) active tags that have their own battery and give power to the system.

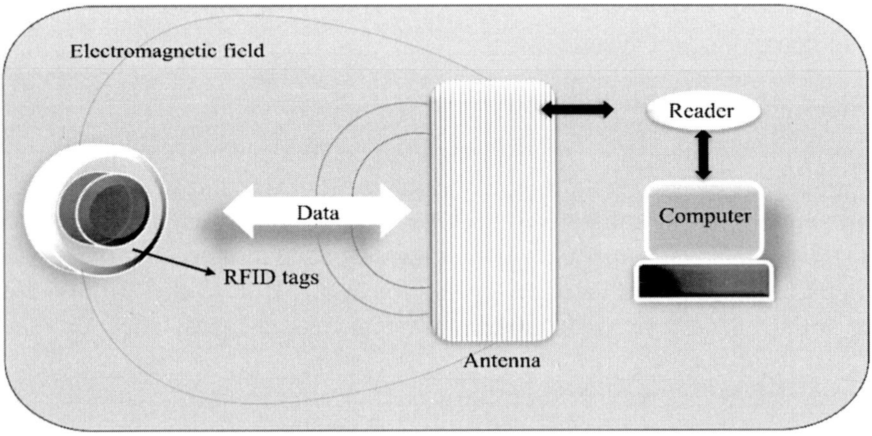

Fig. 10: The working of radio frequency identification (RFID) tag.

8.6. Role of Nanotechnology in Active / Intelligent Packaging

Nanotechnology is the study and use of ultrafine particles having diameter 1-100 nanometer. Due to size in nanometer, these particles are called nanoparticles and show significantly different physical and chemical properties as compared to base macroscale materials. These properties have been heavily exploited to improve the barrier, mechanical and structural properties of base biodegradable and non-biodegradable packaging materials. In recent years, nanoparticles are designed to actively release antimicrobials, antioxidants, enzymes, flavors, nutraceuticals and to actively absorb oxygen, ethylene and UV to improve the shelf life and minimize the quality losses during storage. The main advantages of adopting active nanoparticles are a) very specific applications (if it is designed to release antioxidant, it will not affect other properties); b) very less quantity required as compared to traditional active agents; and c) higher stability under extreme conditions. Some most common nanomaterials explored for food applications are silver, magnesium oxide, copper oxide, nano titanium dioxide and carbon nanotubes. In general. nanoparticles are directly incorporated with the packaging materials which help to improve mechanical and barrier properties and also serve as antioxidant, and antimicrobial agent. The mechanism as antimicrobial agent has not been well understood till now; however, ultrafine nanoparticles stick with cell membrane surface and drastically alter permeability and respiration of the microorganism. They often penetrate through microbial cells and damage them. Many inorganic nanoparticles like silver, copper, magnesium oxide, titanium oxide, and zinc oxide have been found effective in limiting the microbial spoilage. On the other hand, iron, platinum and palladium nanoparticles have been found to reduce the rancidity by scavenging oxygen.

In case of intelligent packaging, nanosensors are quite popular these days and are embedded in food or package to monitor food from production to consumption. Nanosensors are nanoscale devices (tiny chips) that measure physical and chemical changes and convert these to signals that can be detected and analyzed to retrieve valuable information. However, nanosensors have limited area of measurement and it requires network of wireless nanosensors to effectively serve its purpose in intelligent packaging. This network of nanosensors are known as wireless nanosensor network and it has been found very effective in recent years due to significant research and developments. However, nanosensors have limited energy (stored in nano speck) and it requires significant amount of energy to communicate the signal. Thus, scientists around the world are exploring the effective ways to replenish energy stored in nanodevices. In addition to the technical aspects, no regulations on nanosensors applications have been fully established because of lack of sufficient and reliable fundamental researches. Therefore, despites numerous potentials of nanoparticles, use in active and intelligent packaging is still limited.

References

Ahvenainen, R. (2003). Active and intelligent packaging: an introduction. In Novel food packaging techniques (pp. 5-21). Woodhead Publishing.

Barros-Velazquez, J. (Ed.). (2015). Antimicrobial food packaging. Academic Press.

Ghaani, M., Cozzolino, C.A., Castelli, G., & Farris, S. (2016). An overview of the intelligent packaging technologies in the food sector. Trends in Food Science & Technology, 51, 1-11.

John, P. J. (2010). A handbook on food packaging. Daya Publishing House.

Robertson, G. L. (2016). Food packaging: principles and practice. CRC press.

Yildirim, S., Röcker, B., Pettersen, M. K., Nilsen-Nygaard, J., Ayhan, Z., Rutkaite, R., & Coma, V. (2018). Active packaging applications for food. Comprehensive Reviews inFood Science and Food Safety, 17(1), 165-199

9

Modified Atmosphere Packaging

Shivani Desai, Tapas Roy, Dhiraj Kumar Yadav

Department of Food Engineering, National Institute of Food Technology Entrepreneurship and Management, Kundli, Sonipat, Haryana

9.1. Introduction

In a fast-paced world, demand for pre-cooked, fresh-like, and convenient products without preservatives is increasing. This demand has triggered the need for better alternative packaging materials for food products like MAP (Modified Atmosphere Packaging), CAP (Controlled Atmosphere Packaging), edible packaging, biodegradable packaging, etc. Packaging plays a crucial role in transport, distribution, storage, retailing, and end-use, a means of ensuring safe delivery to the ultimate consumer in sound conditions at optimum cost. It performs the functions depicted in Fig.1 for both fresh as well as processed products (Opara et al., 2007; Opara, 2009).

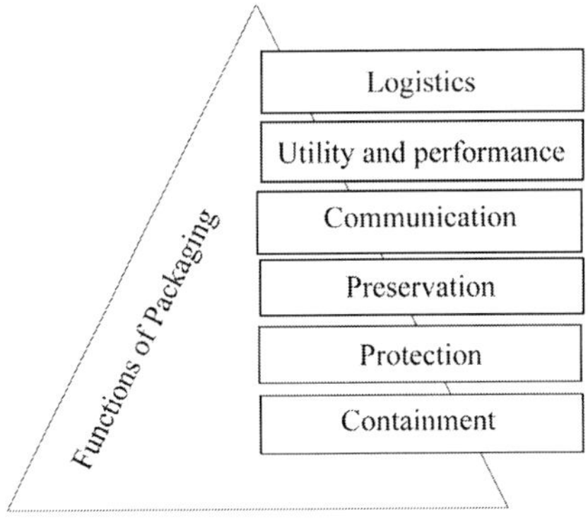

Fig. 1: Functions of packaging

Modified atmosphere packaging (MAP) is the enclosure of fresh and processed food products in gas-barrier packaging materials, in which the gaseous environment has been modified according to the commodity to reduce spoilage and retain the quality for a prolonged period of time. It achieves the composition of gases different from normal atmospheric composition.

Although (MAP) is a relatively new technology, it has a long history dating back to the 1930s. Developments in MAP for various food products are shown in Fig. 2. Just like other packaging techniques, MAP prevents contamination, reduces undesirable physiological, and biochemical changes in food, and reduces microbial growth by altering the surrounding atmosphere.

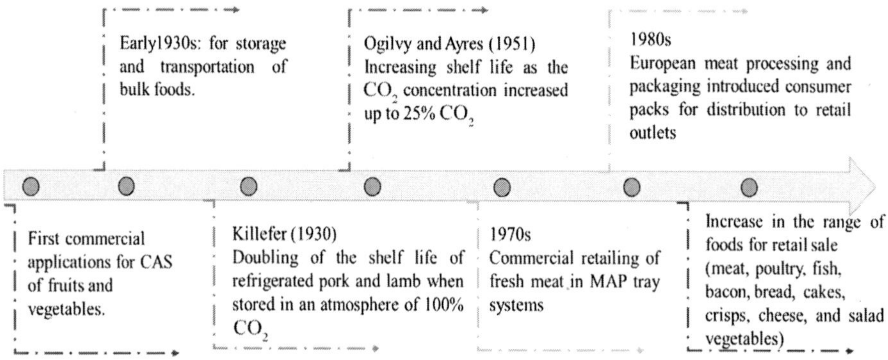

Fig. 2: Developments in MAP (Mullan and McDowell, 2011)

9.2. Advantages and Limitations of MAP

The MAP offers numerous advantages in packaging and safety, but there are certain limitations or disadvantages associated with this which are enlisted below:

Advantages of MAP

- Longer shelf life
- High-quality product
- Improved appearance
- Odorless and convenient packaging
- Good barrier properties
- Can be distributed over a larger distance
- Little or no need for chemical preservatives

Disadvantages of MAP

- Added costs for gases, packaging materials, and machinery
- Different formulations of a gas mixture for different products
- Chances of drip

- Need for temperature control
- Special equipment and training requirements
- Increase in pack volume adversely affect transportation cost and retail display
- CO_2 dissolving into the food could lead to pack collapse and increased drip.

9.3. Principle of MAP

The principle of MAP is the replacement of the surrounding atmosphere of the product within the packaging with an appropriate gas mixture (Sivertsvik et al., 2002) as shown in Fig 3. As the replacement or modification of the gaseous environment is done in MAP, it can be also referred to as'Gas packaging' or 'Gas exchange packaging.' There is no control over the gas composition once gases are introduced in MAP. The food is susceptible to three principle spoilage mechanisms such as oxidation, bacterial growth, and mold growth. All of these mechanisms are suppressed using the appropriate modified atmosphere within the packaging. The removal or addition of gases to control the physicochemical, microbial, and enzymatic activity are safe for both the commodity and the consumer.

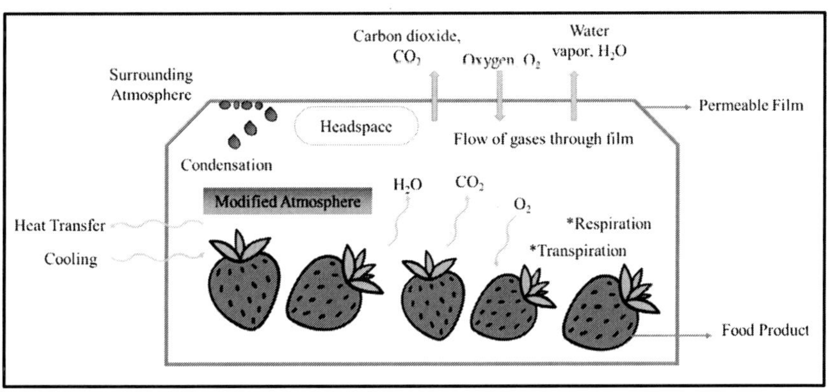

Fig. 3: Modified Atmosphere Packaging

The modified atmosphere normally contains a higher concentration of CO_2 and a lower level of O_2, but it totally depends on the commodity and characteristic of the packaging film. The composition of such gases in MAP can reduce ethylene (C_2H_4) production, respiration rate, ripening, softening, and physiological changes in the product. The principles of preservation and extension of the shelf life by MAP, are shown in the Fig. 4.

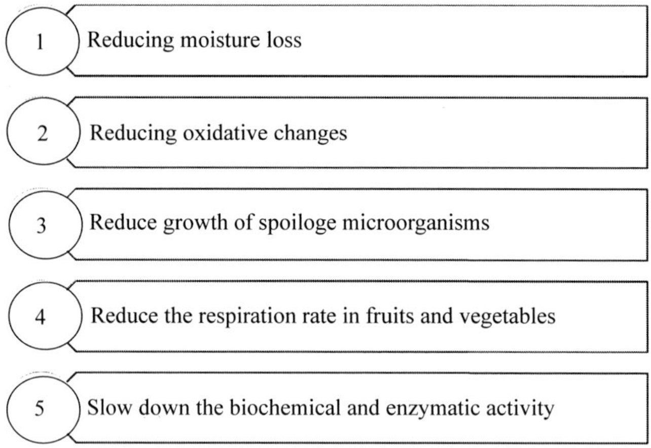

Fig. 4: Principles of preservation by MAP

The effectiveness of MAP depends on various extrinsic and intrinsic factors, viz., type and properties of produce, water activity, pH, gas mixture, type of packaging material, gas to product volume ratio, barrier properties of packaging material, hygiene practices during handling, packaging, storage temperature, and transportation, as shown in Fig. 5 (Opara et al., 2019).

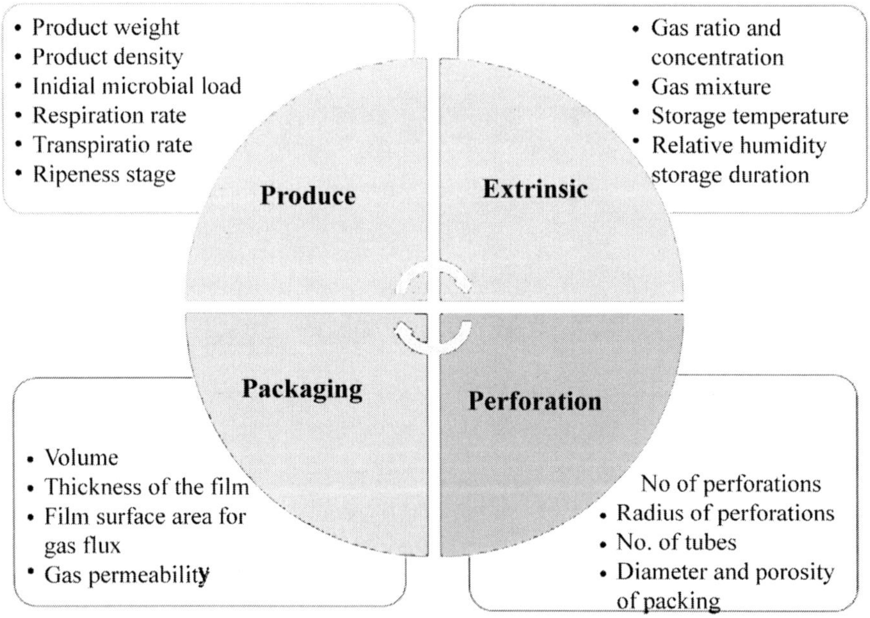

Fig. 5: Factors affecting the effectiveness of MAP

9.4. Gases Used in MAP

The three main gases used in MAP (Fig.6) are oxygen (O_2), carbon dioxide (CO_2), and nitrogen (N_2). The majority of commercial MAP products have employed a mixture of oxygen, nitrogen, and carbon dioxide for product safety, consumer satisfaction, safety issues, cost, etc. Some other gases like helium (He), argon (Ar), and nitrous oxide (N_2O), either alone or in combination, are selected to extend the shelf life and retention of physiochemical and organoleptic properties of food products that are being packed. In general, gases have a low mass (very low density) compared to food (liquid or solid phase), so they disperse homogeneously into the head or void space and can be adsorbed into the food matrix.

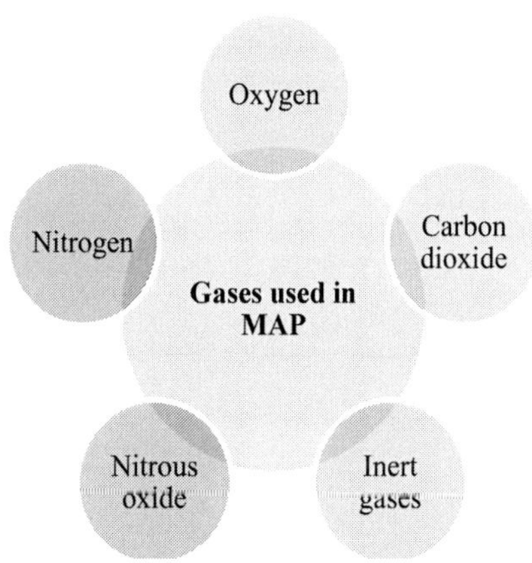

Fig. 6: Gases used in MAP

The different gases used in MAP offer various advantages according to their properties, depicted in table 1.

Table 1: Properties and advantages of gases used in MAP

Gases used in MAP	Properties	Advantages
Oxygen (O_2)	• Colorless, odorless • Highly reactive • Low solubility in water (0.040g/kg at 100kPa, 20 °C) • Promote deteriorative reactions viz., oxidation, browning, and enzymatic oxidation • Required for most bacterial and fungal growth	• Important in meat packaging to retain red color pigment

Gases used in MAP	Properties	Advantages
Carbon dioxide (CO_2)	• Colorless with a slightly pungent odor at a very high concentration • High solubility in water, lipids, and other organic compounds • Solubility increases with decreasing temperature • High permeability • Do not support oxidation, browning, and enzymatic oxidation • Unfavorable for the growth of microorganisms, bacteria, and mold	• Shelf-life extension • Reduction in pH of the food • Reduces oxidative reactions • Affect growth of the gram-negative bacteria
Nitrogen (N_2)	• Inert • Colorless, odorless, tasteless • Low solubility in water and food • Do not support oxidation, browning, and enzymatic oxidation • Do not support the growth of microorganisms, bacteria, and mold	• Extend shelf life • Reduces oxidative reactions • Affect growth of the microorganisms
Noble/inert gases	• Inert • Less reactive	• Balancing gases in MAP • Preserve and extend the shelf life of fresh and minimally processed products

9.4.1. Carbon Dioxide (CO_2)

Carbon dioxide possess some unique properties, i.e., colorless gas, high solubility in aqueous and antimicrobial activity. It is especially effective against gram-negative, aerobic spoilage bacteria that produce meat, poultry, and fish an off-flavor and off-color. Aerobic microbes like Moraxella, Pseudomonas, and molds are generally more sensitive to CO_2 gas than anaerobes like Clostridium and Lactobacillus. Therefore, CO_2 does not inhibit the growth of all microorganisms (Parry, 2012). Because CO_2 is more soluble in water (179.7 ml/100 ml water at 0°C), its inhibitory effect is stronger at lower temperatures. When CO_2 dissolves in water, it produces carbonic acid (H_2CO_3), making it an acidic solution and reducing the pH level of the solution.

$$CO_2 + H_2O \leftrightarrow H_2CO_3\ (aq) \leftrightarrow HCO_3^{-1}\ (aq) + H^{+1}\ (aq)$$

Its high solubility means it requires less headspace and has a direct influence on volume reduction of gas or pressure after initial gas flushing under a higher concentration of CO_2. The antimicorbial activity of CO_2 is significantly greater at 10 °C, as solubility of CO_2 increase with decrease in temperature (Sandhya

2010). Therefore, it has been an effective implication for the MAP of foods. Carbon dioxide gas provides the most effective benefits in MAP application to suppress the rate of microbial growth. CO_2 gas can be utilized at 25 to 100%, depending on its application and type of the food product. The gas mixture containing more than 20% CO_2 has a significant impact on food contamination and many microbial pathogens.

Pack collapse is a common phenomenon in packaging that occurs when items with a high moisture/fat content, like meat, poultry, and shellfish, absorb too much carbon dioxide. Dairy products like creams are extremely sensitive to the high concentration of CO_2 and become contaminated. Therefore, high levels of CO_2 can cause discoloration and acid taste in fish or meat. So, the concentration of CO_2 is an important aspect to increase the shelf-life of food products as well as packaging materials.

9.4.2. Nitrogen (N_2)

Nitrogen is the most commonly used gas in MAP. It is an inert gas with no color, odor and taste, non-flammable and has low solubility in water (0.018 g kg^{-1} at 100 kPa, 20°C). As it is inert in nature and almost insoluble in water and fat, it will not be absorbed in food. N_2 replaces the O_2 in oxygen-sensitive food products to reduce oxidative reactions, which are responsible for rancidity and inhibit the growth of aerobic microorganisms. It is mostly used to inhibit the growth of aerobic microbes, but it does not stop the growth of anaerobic microbes (Mullan & Mcdowell, 2003). With its low solubility in foods, it can be used as a filler gas to prevent shape change or pack collapse when CO_2 or N_2O is used at high concentrations in MAP (Kargwal et al., 2020). Nitrogen gas can be commonly used when carbon dioxide (CO_2), oxygen (O_2), and other gaseous components are not in the right proportion.

9.4.3. Oxygen (O_2)

Oxygen is responsible for oxidative reactions and the growth of aerobic microorganisms. So it is set at a very low level in MAP. The product may be kept in O_2 free environment if the oxidative rancidity is the only parameter responsible for spoilage. High O_2 concentration favors the formation of oxymyoglobin, which is responsible for the red color of meat.

9.4.4. Nitrous oxide (N_2O)

Nitrous oxide (N_2O) is soluble in fat and water and highly permeable into a plastic packaging film. It can be also used to suppress the ethylene production of fresh fruits and vegetables. The oxygen gas is replaced by nitrous oxide to prevent the growth of aerobic microbes and oxidative reactions.

9.4.5. Inert gases

Argon (Ar), helium (He), and xenon (Xe) are common noble or inert gases, which are characterized by their less reactivity. These noble gases are soluble in water and fat and permeable through plastic film. Through plastic permeability range of argon is between oxygen and nitrogen where as permeability range of helium is more than oxygen and less than carbon dioxide. These gases are also applied for tracer gas for detection of leakage of packaging materials. These gases are used in a wide range of food applications, such as potato-based snack products (Sandhya 2010).

9.4.6. Mixture of Gases

The formulation of a gaseous mixture is different for the packaging of different food products, depending upon the product to be packed and desired characteristics, inhibiting the activities of microorganisms to reduce the spoilage.

In modified atmosphere packaging, normally three types of gas mixtures are used normally such as (Parry, 2012):

- **Inert blanketing:** Only with nitrogen (N_2)
- **Semi-reactive blanketing:** With carbon dioxide, nitrogen, or oxygen (CO_2/N_2 or $O_2/CO_2/N_2$)
- **Fully-reactive blanketing:** Carbon dioxide or nitrogen (CO_2 or CO_2/O_2)

The success of fruit preservation in a controlled atmosphere depends mainly on correct selection of highly specific gas mixture (O_2, CO_2, and N_2) for each type of fruit and for different varieties of the same fruit, based not only on biochemical and physiological properties of the fruits but also on the conditions of their cultivation, harvest, and preservation. There are two types of gas regime adopted for the storage of fruits and vegetables: normal and subnormal regimes.

Normal Regimes: Total O_2 and CO_2 content remains same as normal air (i.e. 21%). A mixture of gases containing 5–10% CO_2, 11–16% O_2, and 79% N_2 is most widely accepted. The effect of normal mixtures is based primarily on a high concentration of CO_2, with a threshold value of the order of 10%. In this case, a concentration of O_2, that is reduced minimally to 11% has no appreciable effect on delays of biochemical processes occurring in fruit. Normal regime is considered less effective.

Subnormal Regimes: A mixture of gases with total O_2 and CO_2 content below 21% is used. The most common subnormal mixtures are those with 3–5% O_2, 3–5% CO_2, and 90–94% N_2. This is more common now. It slows the ripening of fruits by depleting the environment of O_2 and enriching it with CO_2.

9.5. Effect of the Gaseous Environment on the Activity of Microorganisms

Microbiological safety is the major concern in food packaging, quality, and safety. The aerobic microorganisms responsible for the food spoilage are inhibited in MAP with moderate to high levels of CO_2 and a very low amount of oxygen. Food contains different types of food spoilage microorganisms, including bacteria, yeast, molds, and virus, therefore, the packaging technologist should be concerned about the growth of pathogenic microorganisms. In this section, the effect of modified gaseous environment on major food spoilage microorganism is discussed.

9.5.1. Effect of CO_2

Carbon dioxide has been recognized to have antibacterial properties. Recent research has shown that CO_2 is more impactful against psychrotrophs and it has the potential to increase the shelf life of stored food products. CO_2 gas highly affects the growth of microorganism like *Pseudomonas*, *Aeromonas*, *E. Coli.*, and *Bacillus* spp., while CO_2 has very little effect on growth of *Enterococcus* spp., *Brochothrix* spp., and *Clostridium* spp. (Mullan &Mcdowell, 2003). The effects of CO_2 depends on the storage temperature and gas concentration. CO_2 concentration near 50% (v/v) and storage temperature of 10°C causes significant effects on microorganisms. A minimum of 20% can be normally used to suppress the growth of bacteria and molds, and the optimal concentration of CO_2 seems to be in the range of 20-30%.

9.5.2. Effect of O_2

Most food spoilage bacteria, yeasts, and molds are aerobic, microaerophiles, and facultative anaerobes and require oxygen to grow. Aerobic bacteria require oxygen for growth, including the gram-negative bacteria such as the Pseudomonas genus. Pathogenic bacteria like *Bacillus cereus*, *Vibrio parahaemolyticus* and *Camplobacter jejuni*. Microaerophiles bacteria can grow at low concentrations of oxygen. These microorganism become dominant in O_2 excluding packaging, where low concentrations of O_2 are present due to the limitation in gas flushing operation and package gas barrier. Facultative organisms such as *Brocothrix thermosphacta* and *Shewanella putrefaciens* can grow in both the presence or absence of O_2, but with a lower concentration of oxygen, growth is usually slower. Anaerobes pathogens, e.g., *Clostridium botulinum* are able to grow without oxygen. Vacuum or nitrogen flushed packaging prevent the growth of aerobic pathogenic bacteria, but anaerobic bacteria can grow in these packaging conditions.

9.5.3. Effect of N_2

Nitrogen gas is used to replace oxygen from MAP. When oxygen is replaced the growth of aerobic or microaerophiles food pathogens gets impeded or stopped. Nitrogen is applied to maintain gas pressure inside the high moisture or fat-containing food package and prevent the collapse of packs because CO_2 is highly soluble in water or fat; these food products are able to absorb CO_2 from the pack, e.g., meat.

9.6. Methods of Creating MA Conditions

Modified atmosphere inside a package can be created either by active or passive method. Passive MAP relies on respiration of the packaged product and the permeability of the packaging film to achieve the desired level of different gas (usually CO_2 and N_2) over time (Oluwafemi et al., 2013). Passive MAP is also called commodity-generated modified atmosphere and is commonly used for packaging fruits and vegetables in modified form. On the other hand, active MAP uses scavengers/absorbers (CO_2, O_2 scavenger, etc.) and is a rapid process of gas displacement to regain the desired gas composition inside the package (Oluwafemi et al. 2013). The difference between active MAP and Passive MAP are summarized in the table 2.

Table 2: Difference between active and passive MAP.

Active MAP	Passive MAP
Modification of the gas composition inside the package by flushing the package headspace with a desired gas mixture	Modification of the gas composition inside a package due to interplay between the product respiration and the package permeability
Intentionally generated MAP	Commodity generated MAP
Shorter equilibrium time (1-2 hours)	Longer equilibrium time (1–2 days to 10–12 days)
It is suitable for cut apples, minimally processed leafy green vegetables, and pomegranate arils.	It is suitable for Mushrooms, carrots, strawberries, spinach, etc.
Extra investment is required for special machinery for gas flushing and packaging.	No extra cost is involved if the package is properly designed.

The rate of respiration of food product and film permeability are substantial to the natural life of the packaged product. The respiration rate is mainly dependent on two major factors such as intrinsic (surface injury, produce maturity level and its cultivar, pre-treatment process, etc.) and extrinsic factors (gas composition, environmental temperature, and storage time). The respiration process in MAP can occur aerobically or anaerobically (oxygen level less than 1%) (Ares et al., 2007). Therefore, the MAP condition should be designed in such a manner that it can control the physiological process (respiration and transpiration) and film permeability. Both the methods are used to inhibit spoilage and improve food products safety. Two techniques are used to modify the pack atmosphere:

9.6.1. Vacuum Modified Atmosphere Packaging (VMAP)

It is a two-stage process. The evacuation stage: The vacuum achieved is generally between 0.6 and 1.3 kPa. Gas flushing stage: the pack is flushed with the modified gas mix. The amount of gas (usually air) removal depends on the amount and duration of gas flushing or application of vacuum before sealing the package (Dobrucka et al., 2015). The gaseous composition can change during storage due to microbial and enzymatic reactions (Nagarajaroa, 2016). There are many factors which can modify the gaseous composition of the package such as temperature, vapor barrier properties of the package, and light (McMillin 2020).

9.6.2. Gas Flush Modified Atmosphere Packaging (GMAP)

This method employs a continuous gas stream that flushes air out from the package prior to sealing. It is less effective at flushing out air. It generally results in residual oxygen levels of 2–5%. Hence, it is not suited for oxygen-sensitive food products. In GMAP, air can be removed either by creating a vacuum or by continuous flushing of gas mixture. In the process of creation of vacuum, initial gas is removed from the package and back flushing of the desired gaseous mixture is done with the help of suitable equipment. While in continuous flushing, the mixture is flushed till the desired amount of gas inside the package is obtained.

9.7. Equipment for MAP

In MAP, equipment are mostly used to create a system for the package with barrier properties to create vacuum or gas flushing conditions. Most VMAP and GMAP are made by either using plastics such as thermoplastic, polytetrafluoroethylene, or polyamide (nylon), or combined with other low-cost and functional type of packaging materials (McMillin 2020). Thermoplastics are used for food products because of their easy moulding and shaping characteristics, recyclability, and being available in different forms such as low density polyethylene (LDPE), polypropylene (PP) and laminated films, etc. Packages for vacuum packaging can be thermoformed (re-melted) from roll-stock film or premade into pouches, whereas packages for gas altered atmosphere packaging can be thermoformed from roll-stock film to form base connections sealed to pre-formed base trays with lidding film (McMillin 2008). An antifogging agent such as polyglycerol eaters, alcohol ethoxylates, glycerol easters, etc. can be used in GMAP. Antifogging agent help in preventing the film fogging by reducing the surface tension of moisture caused by condensation on the internal surface of the film (Bauer et al., 2019).

Mainly, two types of machines are used for MAP, i.e., pillow wrap and chamber type. The basic principle involved in such machines are removal of air by gas flushing from food containing package. Thermoforming or pre-made trays are used in chamber machines. The pouch is positioned in an evacuation chamber in a chamber pouch machine, where a vacuum removes air from the unsealed side (Fig.7). The pouch is heat sealed after the specified time or after certain level of evacuation. Heat is used on the plastic film to form a cavity into which the food is inserted in thermoform and vacuum seal equipment. After the atmospheric air has been removed by evacuation or continuous gas flushing, a second sheet of flat plastic is heat-sealed to a top of the thermoformed package. In order to create pillow-pack packaging, the plastic film is formed into a tube, the two edges are sealed, the food is then placed within the encased plastic tube, as well as the other end of the tube is sealed. Depending on the food type and desired food qualities into the packaged, and evacuation, gas flushing, and/or gas insertion may be performed before sealing.

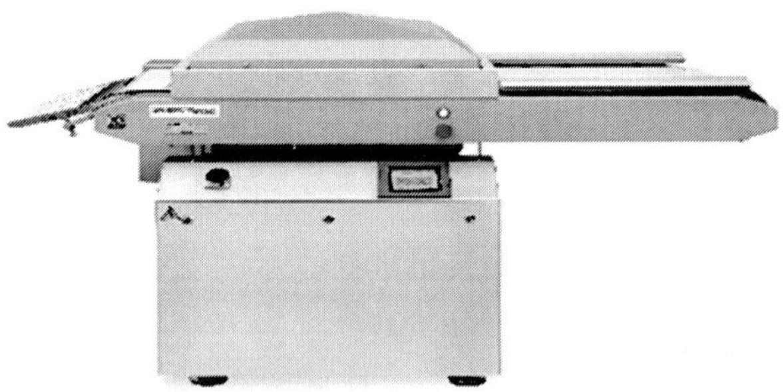

Fig. 7: Chamber pouch packaging machine (McMillin, 2020)

Pillow wrap packaging machine involve horizontal and inverted horizontal form-fill seal and vertical form-fil seal machines. The horizontal form-fill-seal machine (HFFS) (Fig. 8) is used to pack food products in pillow- pack style. It is flexible and uses a roll of polymeric film, the roll is passed through a folding box, also called forming collar. The edges are then sealed together by a heated roller while it is under a specific amount of pressure after being turned into a tube. Atmospheric modification can be done by gas flushing through a lance which is passed-in into the tube of HFFS machine during operation (Arvanitoyannis, 2012).

The packaging material is fed into the inverted horizontal form-fill seal machine (IHFFS) from the bottom, and the top of the pack is brought together and sealed after the two edges are brought together. A lance and the guides that help in the formation of the tube are used to introduce gas. Before the film enters the cross-sealing unit, heated rollers above the tube seal its edges. The sealing quality of package is dependent on temperature, pressure and the contact time, however it varies for different manufacturer because each machines has its own optimum sealing time. The efficiency of the gas flushing unit is also a factor that influences a machine's output. The more quickly a machine runs, the more gas must be pumped into the wrapper tube to displace the air drawn in by the product's passage. When packing crumbly or dusty products, an inverted machine has several advantages. Any debris that falls from the product will remain within the pack which could otherwise cause a poorly sealed and thus leakages in pack. This machine is normally used for MA packaging of sliced meat and cheese.

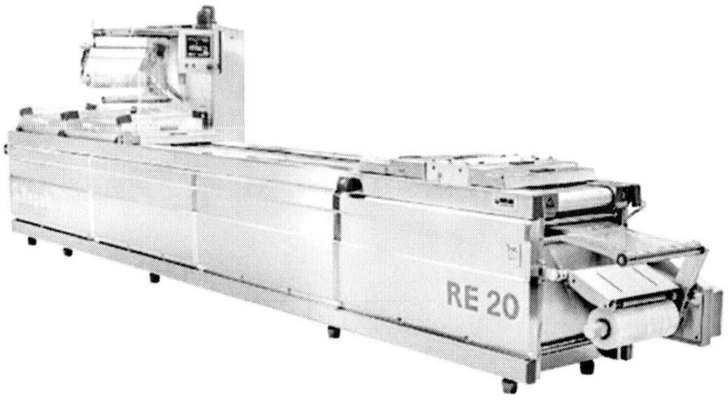

Fig. 8: Horizontal form-fill-seal packaging machine (McMillin, 2020)

A small vertical form-fill-seal packaging machine (VFFS) is depicted in figure 9. After HFFS, a VFFS machines are widely used food industry packaging machine. It involves the common principle as other machines. The only difference is in the method of transportation of packaging materials. A working operation involves pulling a single web of packaging material from a reel, registered it if it is printed, passing it over a forming shoulder, and wrapping it around a metal tube that is fixed in the vertical plane. After being pulled from a reel and registered if printed, the single web of packaging material is then carried over a forming shoulder and wrapped around a metal tube that is positioned in the vertical plane. The tube serves as the folding box, and the two film edges are guided to produce either a fin that is guided via heated rollers or an overlap seam that is sealed by a heated bar pressing down on the seam area. This device is used to package granular and freely flowing commodities like snacks, almonds, and coffee, etc.

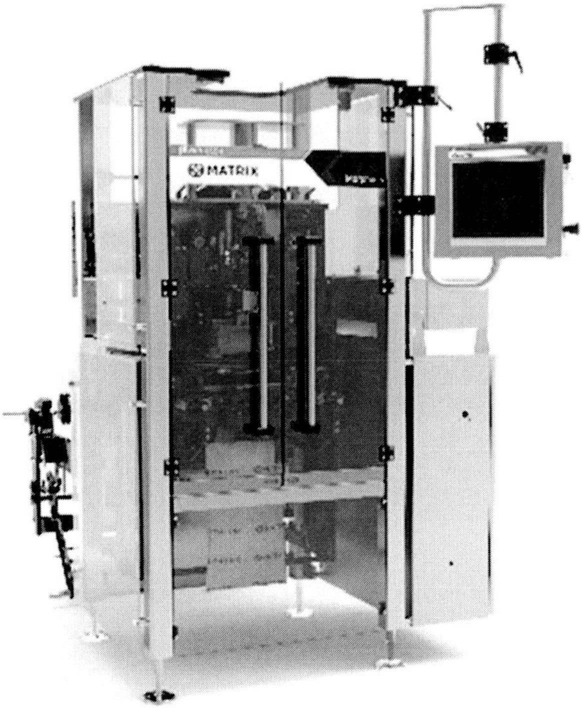

Fig. 9: Illustration of a vertical form-fill-seal packaging machine (McMillin, 2020)

9.8. Packaging Materials for MAP

The highest concern for MAP foods is their safety and quality; thus, choosing the right packaging material is crucial. The most commonly used materials for MAP foods are flexible and semi-rigid plastics and plastic laminates. These plastic materials are useful as food packaging materials due to their relative simplicity of manufacturing, light-weight, good clarity, heat sealing, and strength. Modern polymer processing techniques have made it possible to create polymers that are better suited for specific food packaging applications (McMillin, 2020). However, currently, no single plastic has the characteristics that make it appropriate for all applications in food packaging.

Although some plastic packaging materials are monolayer structures made from a single plastic, most, if not all, MAP films are multilayer structures made up of many layers of a variety of polymers which are further merged using different processes such as coextrusion, laminating, and coating to develop a best suitable packaging material with its best performance to create rigid packs, flexible sheets, and plastic films.

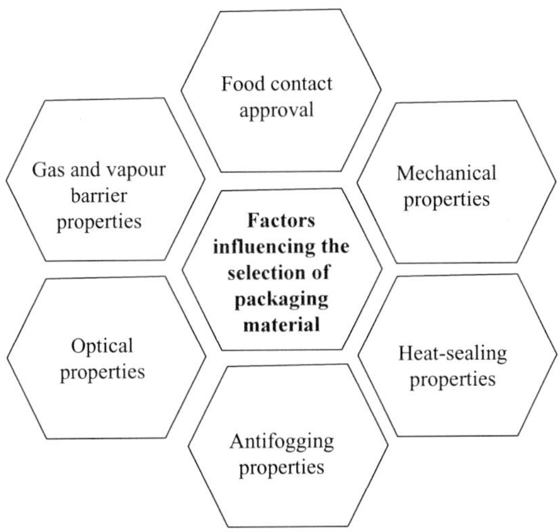

Fig. 10: Packaging material selection parameters for MAP

Factors influencing the selection of packaging material for MAP (Fig.10):

(i) Food contact approval

- Ensuring that the packaging material which comes in direct contact of food must not be a source of harmful contaminants and provide adequate protection against other form of contaminants.

(ii) Gas and vapour barrier properties

- Optimal gas and vapour barrier properties of packaging material ensures the quality and safety of food, along with prevention from external contaminants.

(iii) Optical properties

- Packaging material with high transparency and glossy nature are preferred for satisfying the customers demand for clear view of the product.
- Colored base trays are also used to enhance the visual appeal of the products.

- For eg: PVC trays are preferred for their transparency and PET, PP and PET for colored trays.

(iv) Antifogging Properties

- Condensation of water vapour on the inner surface of the packaging material is caused due to the difference in the temperature which affects the visibility and aesthetic appeal of the product.
- Applying antifogging agent such as fatty acid ester, as inner additives or external coatings to heat sealing layer can prevent the moisture to form larger vapour droplets.

(v) Mechanical properties

- Material resistant to tearing and puncture, and good machine handling properties, maintains the integrity of the packaging during its handling and distribution.

(vi) Heat sealing properties

- To maintain the desired gas composition inside the pack, effective heat seals are required which depends on heat seal material, seal width and machine settings such as temperature, pressure and dwell time.

Most frequently, flexible films are used for making bags, pouches, pillow packs, and top webs in MAP applications. These flexible films can be produced from Polyethylene (PE), Polypropylene (PP), Polyamide (nylons), Polyethylene Terephthalate (PET), Polyvinyl Chloride (PVC), Polyvinylidene Chloride (PVDC), and Ethylene Vinyl Alcohol. Along with that, certain rigid and semi-rigid constructions are used for making foundation trays, dishes, cups, and tubs. These can be made using PP, PET, unplasticized PVC, and expanded polystyrene. To meet the desired film characteristics for MAP, the different plastic films are either laminated or coextruded. Most commonly used plastics used in MAP and their effects are discussed below:

9.8.1. Ethylene Vinyl Alcohol

Poly vinyl alcohol (PVOH) is copolymerized with ethylene to enhance the barrier properties of the material. PVOH itself has good barrier properties in its dry state, but it is exposed to water or moisture, it retains the moisture in itself and swells the plastic making it plasticized. For the purpose of commercial use and making a better material, even in the presence of moisture or water, it gets copolymerized with ethylene which not only makes it a better material for MAP applications but also has less sensitivity to water or moisture.

Additionally, EVOH has great mechanical strength, strong oil and organic solvent resistance, and high thermal stability (Marsh and Bugusu 2007).

9.8.2. Polyethylene (PE)

The most conventional plastic material for packaging purposes is polyethylene, which is structurally the simplest category of synthetic polymers. All of these are made of a carbon backbone with a degree of branched sidechains which makes variations in their density. Based on the density of the plastic material, it is classified into different forms, i.e., Low-density polyethylene (density, 0.910-0.925 g cm^{-3}), High-density polyethylene (density, 0.940 g cm^{-3}), commonly used for rigid and semi-rigid structure (Bodbodak and Moshfeghifar 2016). PEs have high hydrophobicity, which makes them efficient against water vapor. But for MAP applications that require high barrier properties, it itself cannot be used because of its melting property at low temperatures ranging from 100-120°C. In order to impart heat-sealing property to base trays and lidding films, a sealant layer called linear low-density polyethylene (LLDPE) is used because of its high heat-sealing characteristic properties.

9.8.3. Polypropylene

A versatile polymer with wide applications in flexible, rigid, and semi-rigid packaging structures and has the lowest density among commodity plastics. It has good resistance to chemicals and effective water vapor barrier properties but not desirable gas barrier properties. This gas barrier property can be somewhat enhanced by increasing the thickness of the material. Although it has a high melting point which makes it suitable for thermal resistance applications in MAP, laminated with other resistant materials for making thermoformed base trays to provide heat sealing properties (Mangaraj et al., 2009).

9.8.4. Polyamides

Polyamides primarily comprise nylons, which have wide applications in the field of food packaging. They often have strong tensile characteristics, good abrasion and puncture resistance, and effective gas barrier qualities, as well as great aroma and flavor barriers, but frequently exhibit a tendency to shed some barrier efficacy when exposed to significant levels of moisture and are extremely poor water vapor barriers. Their relatively strong strength and agility make them perfect for fresh meat vacuum pouches where hard bone ends might pierce other plastic substances. They are essentially coextruded with other materials to impart some distinguishing qualities because they are unsuited for commercial purposes and have significant production costs. The most popular nylon packaging film in the industry is typically nylon 6.

9.8.5. Polyethylene Terephthalate (PET)

The most prevalent polyester used in food packaging applications is polyethylene terephthalate. PET is robust, has good clarity, and is temperature resistant along with favorable gas and water vapor barrier properties. By applying PVDC to the coating, the barrier qualities can be improved. Good thermal seal qualities are frequently provided through coating or co-extrusion. For short durations, it can withstand much higher temperatures, like in the case of dual ovenable packaging for frozen goods. Its glass-like transparency, appropriate gas barrier qualities, light weight, and shatter resistance are the primary factors in its acceptance in the food packaging industry (Kirklanda BS et al. 2008).

9.8.6. Polyvinyl Chloride (PVC)

PVC is formed by adding PVC with plasticizers and other additives to make it a flexible film. In general, these films have good self-cling, toughness, medium strength, outstanding chemical resistance, resilience, and clarity. Unplasticized PVC has enhanced gas and water vapor barrier qualities than plasticized ones. Although it has excellent resistance to oil and grease, some organic solvents can soften it. In order to offer the necessary heat-sealing qualities, it is laminated to PE and used as a structural material in MAP thermoformed base trays.

9.8.7. Polyvinylidene Chloride (PVDC)

PVDC is a copolymer of vinyl chloride and vinylidene chloride possessing excellent heat sealing, odour barrier, and O_2 and CO_2 barrier properties They are not significantly affected by the presence of moisture. The packaging of poultry, cured meats, cheese, snacks, tea, coffee, confectionery, some of the key applications of PVDC.

9.9. Applications of MAP

MAP can be used to actively or inertly change the environment around the food product inside the package. In order to create the best conditions for extending the shelf life and reducing oxidation and food spoilage as well as the deterioration of perishable products, this modified atmosphere is specifically created by altering the natural composition and distribution of atmospheric gases within packaging.

MAP is frequently used to safeguard fresh food by reducing oxygen exposure and, consequently, oxidation. Oxidation can cause discoloration, deterioration, and changes to flavours and textures to food products. Applications of MAP in different sectors of food (Fig. 11) are reported in following sections.

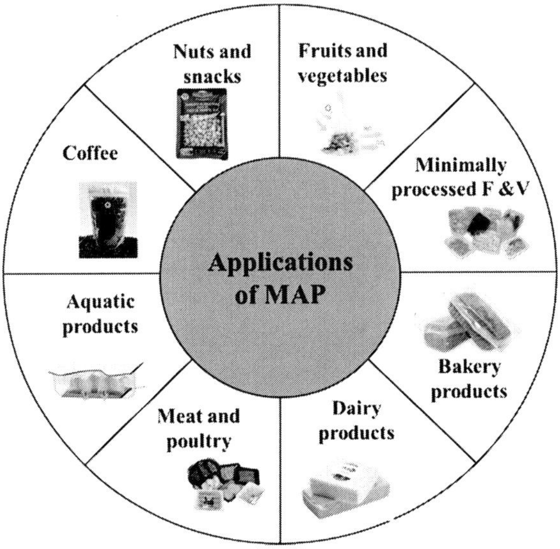

Fig. 11: Applications of MAP

9.9.1. Agricultural Products including Fresh Fruits and Vegetables

MAP in fruits and vegetables aims to reduce the respiration rate of the product.This greatly reduces the production of ethylene, a gas that hastens the development of senescence in fruits and vegetables and speeds up the ripening and deterioration of produce. Also, they decrease the softening rates, and improves chlorophyll and other pigment retention. In addition, increased CO_2 level inhibits the microbial growth and hence, prevents spoilage.

9.9.2. Snacks and Nuts

A significant portion of the fat in snacks and nuts causes them to oxidise and become rancid. Therefore, in order to avoid spoilage and hence increase shelf life, packaging materials for foodstuffs should be airtight and oxygen levels should be maintained to a minimum. The type and quantity of gases used are entirely dependent on the food product. MAP oxygen can be swapped out for 100% CO_2 or, in some circumstances, pure nitrogen can also be used for packaging of snacks and nuts (Shayanfar et al., 2011). Such packaging materials give both snacks and nuts a longer shelf life and a good barrier quality to moisture and gases.

9.9.3. Meat Products Including Sausage and Poultry

Meat products are more prone to spoilage because of the growth of microorganisms, higher amount of moisture and nutritional components.

Considering high hygiene and quality standards, MAP atmospheres are deemed appropriate for meat and poultry products for enhancing the shelf life. An optimal mixture of gases (O_2, CO_2 and N_2) is preferred for the packing of fresh meat and related products which results in prolonged shelf life by inhibiting microbial growth on fresh food commodities.

Poultry is also highly susceptible to spoilage. For this too, CO_2 atmosphere is used for extending shelf life but with a limited amount as higher levels can cause sour taste to food products. A high level of oxygen is used as for chicken without skin to preserve the color of the meat. Nitrogen is used as balancing gas to prevent collapsing of the packaging material (Czerwiński et al. 2021).

9.9.4. Aquatic Products like Fish and Seafoods

Fish and shellfish are highly perishable, due to their high water activity a_w, neutral pH and the presence of autolytic enzymes which cause the rapid development of undesirable odours and flavors. Some of them also contains fatty acids which causes rancidity due to oxidation. Inhibition and enhancing the shelf life could be obtained through the use of low storage temperatures combined with high CO_2 contents in the atmosphere surrounding the product by MAP. This CO_2 can sometimes cause undesirable effects such as sour tastes and dry products, overcoming this makes use of O_2 as a mixture with CO_2 with appropriate concentrations. This practice can help in prevention of spoilage of aquatic products and extension of shelf-life. For raw fishes an increase of 50-100% in storage life and for cooked a shelf-life extension of 100-200% can be obtained under ideal storage conditions in MAP (Sivertsvik et al., 2002).

9.9.5. Coffee

It is pretty evident that some carbon dioxide is emitted when coffee is being roasted. When they are packaged after being allowed to cool, the CO_2 they produce could cause the packaging material to collapse, impairing the freshness of the beans and producing dampness. Therefore, the right combination of MAP gases can be utilized to inhibit microbial growth and prevent the loss of tastes and odours. It also prevents oxidation of fatty acids which is the major cause of rancidity. For the packaging of capsules and sachets, sometimes, pure nitrogen is preferred for the prevention of microbial, enzymatic physical deterioration and retained freshness of the product (Czerwiński et al., 2021).

9.9.6. Bakery Products such as Cakes and Bread

The primary cause of bakery product shelf-life reduction is microbiological spoiling, particularly mould growth, which results in large monetary losses. The high CO_2 MAP has a very good bacteriostatic and fungistatic property along with inhibitory effect on growth which results in the prevention of mold growth and further leads to shelf-life extension of bakery products. Further for preventing rancidity, another supporting gas is used as a filler gas with low water activity and for replacement of oxygen (Czerwi ´et al., 2021).

9.9.7. Ready Made Meals and Pasta

The ready-to-eat food products consists of a variety of ingredients which have different compositions with different shelf-life and thus, different spoilage parameters. In the case of such products a mixture of CO_2 and N_2 is given the preference with minimal level of O_2. This mixture of gases is totally dependent on the composition of the food products and its packaging. The resultant atmosphere not only delays microbial spoilage but also rancidity which effects the taste of the food because of oxidation (Czerwiński et al., 2021). Such products are sometimes kept away from light to prevent discoloration by packing them in colored prints, pigmentation or, metallized forms packaging materials.

9.9.8. Wine and Drinks

Wine manufacturing uses gases to preserve the quality of the finished product, mainly consisting of argon and food-grade nitrogen. They primarily serve to remove oxygen, and further prevents oxidation and lessens its impact on wine's flavour. Basically, Argon and Nitrogen are used for the production of wine but because of the cost- effectiveness and easy availability of Nitrogen it is being preferred. Nitrogen can be used for a variety of purposes including nitrogen flushing, sparging, and blanketing during winemaking process, all of which assist keep oxygen out of the wine while it is being produced (Sandhya 2010).

9.9.9. Dairy Products

The major spoilage in the dairy products such as cheese is due to the growth of microorganisms and rancidity which can be overcomed by uninterrupted cooling to the products which contributes to its shelf-life extension. It is incredibly simple and effective to obviate the growth of mold on food products by controlling the levels of CO_2 in the packaging.

Certain products require specified quantities of CO_2, which is also compatible with the packaging material to reduce the risk of collapsing (Sandhya 2010). For ex: Soft Cheese requires lower levels of CO_2 to prevent collapsing of the packaging and development of sour flavors to products such as yoghurts, cream, and Milk.

9.10. Recommended MAP Conditions for Various Products

MAP can be suitable for both respiring (meat, fish, poultry, fruits and vegetables) and non-respiring foods (processed food like bakery, coffee, snack food etc.). The recommended gas mixtures for various food products are shown in Table 3. Fruits and vegetables are important perishable commodities, which requires proper storage conditions during the storage and transit. Therefore, to delay the metabolic processes responsible for the spoilage and reduction in shelf-life, MAP plays crucial role. MAP for fresh and minimally processed fruits and vegetables maintains the quality of product as well as extends the shelf life. The recommended gas mixtures for various fresh-cut fruits and vegetables are summarized in Table 4.

Table 3: Recommended gas mixture of MAP (Parry, 2012).

Products	% Oxygen	% Carbon dioxide	% Nitrogen
Red meat	60-85	15-40	-
Cured meat	0	20-25	50-80
Fish (white)	30	40	30
Eggs	0	20	80
Poultry	0	25-30	70-75
pork	80	20	0
Hard cheese	0	100	-
Bread	-	60-70	30-40
Non-dairy cakes	-	60	40
Dairy Cakes	-	-	100
Sandwiches	0-10	20-100	0-100
Pasta	-	-	100
Fruits and vegetables	3-5	3-5	85-95
Bakery	0	20-70	20-80
Dried foods	-	-	100

Table 4: Recommended MAP conditions for various fresh-cut vegetables and fruits (Oliveira et al., 2015).

Products	Temperature (°C)	Atmosphere	
		O_2 (%)	CO_2 (%)
Fresh-cut vegetables			
Broccoli	0-5	2-3	6-7
Shredded cabbage	0-5	5-7.5	15
Sliced leek	0-5	5	5
Chopped butterhead lettuce	0-5	1-3	5-10
Chopped green leaf lettuce	0-5	0.5-3	5-10
Sliced mushrooms	0-5	3	10
Sliced or diced onion	0-5	2-5	10-15
Sliced or whole-peeled potato	0-5	1-3	6-9

Products	Temperature (°C)	Atmosphere	
		O_2 (%)	CO_2 (%)
Sliced rutabaga	0-5	5	5
Cleaned spinach	0-5	0.8-3	8-10
Sliced tomato	0-5	3	3
Fresh-cut fruits			
Sliced apple	0-5	<1	-
Cubed cantaloupe	0-5	3-5	6-15
Sliced kiwifruit	0-5	2-4	5-10
Sliced orange	0-5	14-21	7-10
Sliced peach	0	1-2	5-12
Sliced pear	0-5	0.5	<10
Sliced persimmon	0-5	2	12
Sliced strawberry	0-5	1-2	5-10

9.11. MAP Modeling

The design or selection of packaging material is an important parameter in MAP to maintain the desired gas concentrations. The development of predictive model streamlines and accelerates the design process. The key parameter for developing a predictive model for MAP is the relation between the rate of respiration and the rate of permeation of respiratory gases through the package. The equilibrium is reached when the O_2 consumption rate by the commodity is equal to the CO_2 permeation rate leaving the package. Factors to be considered for predictive models for equilibrium gas concentrations are shown in Fig. 12 (Rao, 2015).

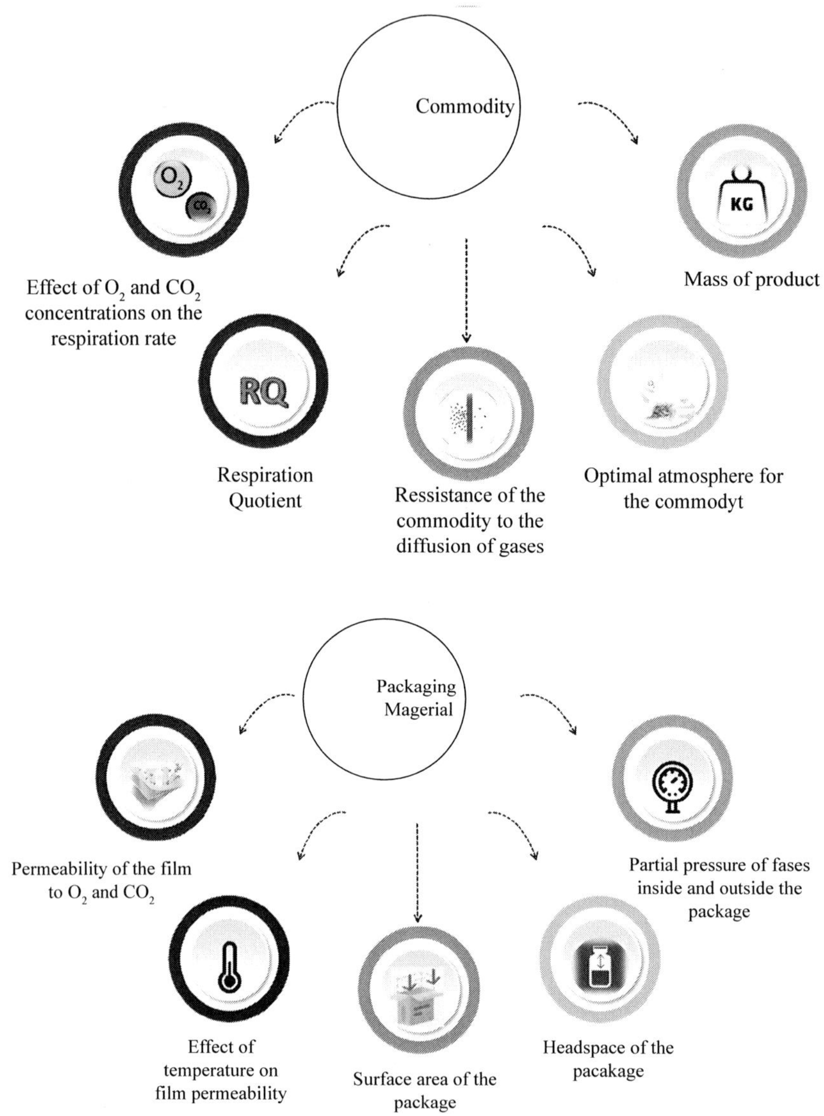

Fig. 12: Factors to be considered for predictive models for equilibrium gas concentrations

Modeling of MAP mainly represents the rate of transportation of components (mixture of gases) present inside the MAP. During the modification of the atmosphere inside the package, a particular concentration of gases has been maintained to increase the shelf life of food by reducing the transpiration rate in live foods and reducing the chemical and enzymatic reactions in processed foods. If we maintain a particular concentration of O_2 and CO_2 in the package or chamber, then the concentration gradient will be the driving force allowing to move the components from higher to lower concentration, and the rate of movement/ diffusion will depend upon the permeability of the package for a particular component.

The influx of O_2 and efflux of CO_2 are determined by the permeability of the package, the thickness and surface area of the film, and the partial pressure gradients of O_2 and CO_2 inside and outside the package at a specific temperature (Robertson, 2005).

9.11.1. Determination of Diffusion of Gases through Packaging Film

The permeability of packaging material is the rate of penetration of gases through the packaging material. It is the mass of gases or vapor transferred per unit of driving force, time, and area.

Fick's Law of diffusion for a component A:

$$N_A = -\Delta_{AB} \times \frac{\partial C_A}{\partial C}$$

Or

$$N_A = -\Delta_{AB} \frac{(C_{A2} - C_{A1})}{(\tau_2 - \tau_1)} \quad (1)$$

Where,

N_A = Diffusion flux of gas A (Diffusion of gases through packaging material)

Δ_{AB} = Diffusion coefficient or mass diffusivity

C_{A2} and C_{A1} = Concentration of A_2 and A_1, respectively

τ_2 - τ_1 = Thickness of film

(The negative sign indicates that the diffusion takes place in the direction of decreasing concentration)

The solubility of solute gas (A) in solid (package film) is usually expressed as 'S' in m^3 solute per m^3 solid per atm pressure of (A), then the concentration of A_1 is C_A in the solid in kg.mol.A/m^3

$$C_A = \frac{S.\left(\frac{m^3}{m^3 \text{solid. atm}}\right)}{22.414\left(\frac{m^3}{\text{kg mol. A}}\right)} P_A\,(\text{atm})$$

$$C_A = \frac{S.P_A\left(\text{kg mol of A}\right)}{22.414(\text{m}^3.\text{solid})} \tag{2}$$

$$C_A = \frac{S.P_A}{22414}.\frac{\text{g mol A}}{\text{cm}^3\text{solid}}$$

Molar volume of an ideal gas at 1atm pressure is 22.414 m^3/kg solid.

Replacing the value of C_A in equation 1 from equation 2,

$$N_A = \frac{\Delta_{AB}.S\left(P_{A1} - P_{A2}\right)}{22.414(\tau_2 - \tau_1)} \tag{3}$$

$\Delta_{AB.}$ S = P_m (Permeability)

Hence,

$$N_A = \frac{P_m\left(P_{A1} - P_{A2}\right)}{\left(\tau_2 - \tau_1\right)22.414} \tag{4}$$

$$N_A = \frac{P_m\left(P_{A1} - P_{A2}\right)}{\varkappa.22.414} \tag{5}$$

If the packaging made of n no. of layers, then,

$$N_A = \frac{\left(P_{A1} - P_{A2}\right)}{22.414.\left(\frac{\varkappa_1}{P_{m1}} + \frac{\varkappa_2}{P_{m2}} + \ldots\right)} \tag{6}$$

Here,

$P_{A1} - P_{A2}$ = Overall Partial Pressure

$\varkappa$ = Thickness of packaging

P_m = Permeability

N_A= Diffusion flux of A or Rate of diffusion of A

Example 1

Polyethene film, 0.00015 m thick is being used in packaging of a product at 80°C. If the partial pressure of O_2 outside the package is 0.21 atm and inside is 0.01 atm. Calculate the rate of transportation of O_2 at steady state.

Solution

$P_m = 4.17 \times 10^{-12}$ .m^3/s.m^2 atm/m

$$N_A = \frac{P_m.(P_{A1} - P_{A2})}{22.414.\varkappa}$$

$$= \frac{4.17 \times 10^{-12}(0.21 - 0.01)}{22.414 \times (0.00015 - 0)}$$

$N_A = 2.480 \times 10^{-10}$ kg mol/s. m^2

Example 2

A 180 μm, LDPE film is being used to wrap fruits in MAP at 28°C temperature. The partial pressures of O_2outside and inside the package are 21.03 kPa and 1.01 kPa respectively, whereas the partial pressures of N_2 outside and inside the package are 77.98 kPa and 1.02 kPa. Calculate the gas flux through LDPE package at steady state.

Permeability of O_2 @ 28°C = 4.17×10^{-16} m^3 solute at STP m^{-1} s^{-1} KPa^{-1}

Permeability of N_2 @ *28°C* = 12×10^{-12} m^3 solute at STP m^{-1} s^{-1} KPa^{-1}

Solution

Thickness of film: 180 μm

Outside partial pressure of O_2 = 21.03 kPa

Inside partial pressure of O_2 = 1.01 kPa

Outside partial pressure of N_2 = 77.98 kPa

Inside partial pressure of N_2 = 1.02 kPa

Flux of Oxygen

$$N_A = \frac{P_m.(P_{A1} - P_{A2})}{22.414.(\varkappa)}$$

Substitute the values in above equation,

$$N_A = \frac{\text{ooooo}\times\text{oo}^{-16} \times (\quad - \quad)}{22.414 \times (180 \times 10^{-6})}$$

$= 2.07 \times 10^{-12}$ kg mol m^{-2} s^{-1}

Flux of Nitrogen

$$N_A = \frac{P_m.(P_{A1} - P_{A2})}{22.414.(\chi)}$$

$$N_A = \frac{1.12 \times 10^{-12} \times (78 - 1.02)}{22.414 \times (180 \times 10^{-6})}$$

$= 2.14 \times 10^{-8}$ = Kg mol m^{-2} s^{-1}

9.11.2. Determination of Surface Area of Permeable Film

The exchange of gases through modified atmosphere packaging takes place due to diffusion. To calculate the exchange of gases or diffusion rate surface area of packaging material is the key parameter. It can be calculated by the following equations (Rao, 2015):

$$F = \frac{R_t\left(0.365 + 2.774C^{i}_{O_2} - 0.883C^{i}_{CO_2}\right) \times m}{P_c\left(C^{i}_{O_2}P^{o} - C^{i}_{O_2}P^{i}\right)}$$

Or

$$F = \frac{R_t\left(0.365 + 2.774C^{i}_{O_2} - 0.883C^{i}_{CO_2}\right) \times m}{P_c\left(C^{o}_{O_2}P^{o} - C^{i}_{O_2}P^{i}\right)}$$

Where,

F = Area of the selectively permeable packaging material, m^2

R_t = Absorption of O_2 and release of CO_2 during respiration, m^3/ton. h

$C^i\ O_2$ = Volume concentration of O_2 (taken as unity)

$C^i\ CO_2$ = Volume concentration of CO_2 (taken as the unity)

m = Mass of the product, ton

P_C = Permeability coefficient for CO_2

Pi and Po = Pressure inside and outside package, respectively, mgHg

Example 3

35 kg Apples with respiration rate of 40 mg/kg·h, are to be stored at (25–30°C) in modified atmosphere package in terms of CO_2 release. The permeability value of the selective package is 7.1×10^{-3} m/ day·atm. The packaging material used to store apples is 70-micron thick LDPE with Po/Pc ratio 0.24. The gas composition inside the package is 2% and 5% respectively. What will be the surface area of selected LDPE?

(The density of CO_2 at 25°C is 1.808 kg m^{-3})

Solution

Rate of respiration, $R_t = \frac{40}{10^6} \times \frac{35}{1000} \times h$

$= 1400 \times 10^{-9}$ kg ton h

$= \frac{1400 \times 10^{-9}}{1.808} \ m^3 / tonh$

$= 774.336 \times 10^{-9}$ m^3/ton h

Surface area of package is calculated by the following equation,

$$F = \frac{R_t \left(0.365 + 2.774 C^i_{O_2} - 0.883 C^i_{CO_2}\right) \times m}{P_c \left(C^i_{O_2} P^o - C^i_{O_2} P^i\right)}$$

$$F = \frac{774.336 \times 10^{-9} \left(0.365 + 2.774 \times 0.02 - 0.883 \times 0.05\right) \times 0.035}{7.1 \times 10^{-3} \left(0.21 \times 0.76 - 0\right)}$$

$$= \frac{774.336 \times 10^{-9} \left(0.365 + 0.05488 - 0.04165\right) \times 0.035}{7.1 \times 10^{-3} \left(0.1596 - 0\right)}$$

$$= \frac{774.336 \times 10^{-9} \times 0.37823 \times 35}{7.1 \times 0.1596} = 9.05 \ m^2$$

9.11.3. Graphical Determination of Suitable Packaging Film

The suitable packaging film for the respective commodity can be selected using the graphical relation between rate of respiration and rate of diffusion as shown in the fig 13. The graphical representation of respiration curve shows the relationship between O_2 consumption (cm^3 /bag day) and partial pressure of O_2 inside the package on the y-axis and x-axis respectively. The diffusion curve shows the rate of diffusion, and it is depending on the differences in partial pressures inside and outside the package and the thickness of film.The diffusion curves for different thickness of LDPE package depicts that the higher thickness has lower diffusion rate. Both the curve shows higher diffusion rate in the beginning and then gradually decline.

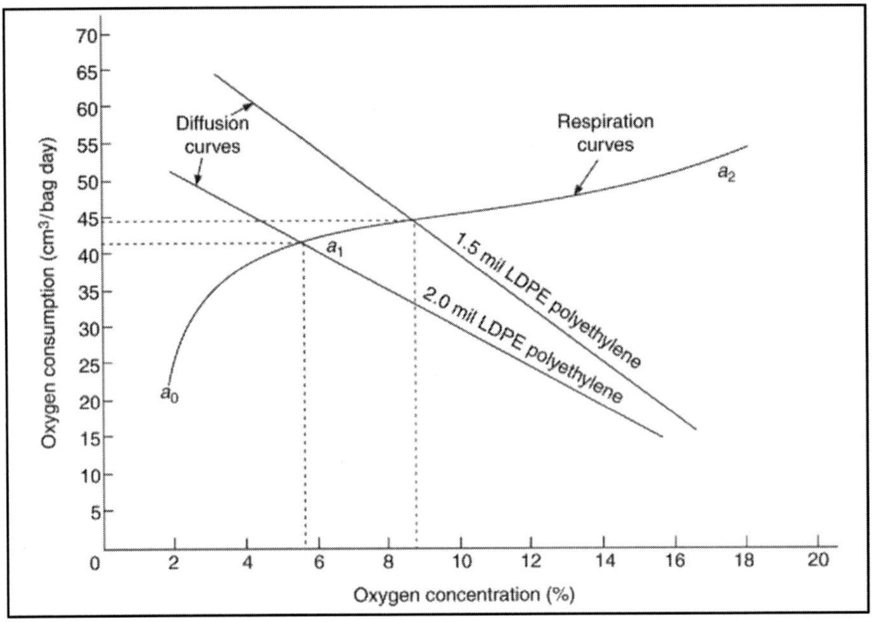

Fig. 13: Graphical determination of suitable packaging material

The point of intersection of both curves (respiration curve and diffusion curve) depicts the equilibrium condition i.e., gas flux due to respiration of commodity which is equal to the gas flux due to permeation of respiratory gases through package. Therefore, the packaging film with diffusion rate corresponding to the intersecting part with the respiration curve should be selected.

9.12. Effect of the MAP on Chemical and Physical Properties of Food

The chemical reactions and enzyme-catalysed biochemical reactions occurring in food are another major cause of food spoilage. These are the main considerations that the industry must make in order to design appropriate packaging for food products to prevent various enzymatic and oxidative reactions occurring inside because of the release of certain organic substances from the food commodity (McMillin 2020).

9.12.1. Effects of Oxygen

The majority of the alterations were brought about by the presence of oxygen in the atmosphere; meanwhile, CO_2 is investigated to a lesser extent. It is commonly known that the presence of oxygen (O_2) encourages the growth of microbes, but it can also encourage lipid oxidation, which alters the flavour and color of food pigments and contributes to enzymatic browning. So overall, it has both good and bad impacts on the properties of food.

Oxidation of Lipids

The lipid or fat oxidation also termed as oxidative rancidity is one of the major causes of food spoilage. When oxygen reacts with unsaturated fatty acids in fat-containing food, it starts its spoilage mechanism called auto oxidation, forming hydroperoxides. Along with this, acids and aldehydes are largely responsible for the off-flavor and off-odor characteristics of rancid foods. In order to overcome this, oxygen can be replaced by Nitrogen (N_2) and Carbon Dioxide (CO_2).

Pigment colour in meat

Meat comprises of three major pigments known, Oxymyoglobin (Red), Metmyoglobin (Brown), and Myoglobin (Purple). The concentration of O_2 present in the packaging atmosphere can lead to certain changes including conversion of myoglobin to oxymyoglobin (presence of high O_2) or metmyoglobin (presence of low levels of O_2) because of the oxidative reactions which ultimately results in the change of color of fresh meat. This formation and reformation of the pigments in meat is dynamic and constant.

Photo-oxidation of chlorophyll

When green fruits and vegetables are exposed to light, it causes photo-oxidation and results in the change in color of the chlorophyll from green to brown/grey when oxidized to pheophytin. This can be significantly reduced by MAP by lowering O_2 levels and storage in opaque packages.

Oxidative off-flavors

The numerous oxidative processes that occur in food and drink products might result in oxidative off-flavours. Oxidative warmed-over flavour is a distinctive off-flavour that is mostly associated to cooked meat and poultry. Commercially, it has an impact on cook-chill products like chilled ready meals. This stale, oxidised flavour may become detectable in cooked meats and poultry maintained at chilled storage temperatures in a short period of time.

Meat, fish, poultry, beverages, and dairy products are particularly vulnerable to oxidative processes, which can start a cascade of events that compromises flavour. This can happen rather quickly. The low O_2 levels in MAP can prevent the emergence of oxidative off flavours.

9.12.2. Effects of other gases in MAP

Since nitrogen is an inert gas, it does not participate in any chemical or biological reactions and has little to no impact on food spoilage. CO_2 being highly soluble reacts with water and forms carbonic acid which can have some adverse impact on particular food products because of the exposure of the low pH areas on or near the food surface. This results in the pH-induced protein changes leading to denaturation and some other confirmational changes.

Carbon monoxide can react with myoglobin and form carboxymyoglobin (Red color compound) similar to oxymyoglobin resulting in a better stability than the existing compound which makes carbon monoxide toxic. Therefore, CO is not one of the gases that is recommended for use in the packaging of food products.

References

Ares, G., Lareo, C., &Lema, P. (2007). Modified atmosphere packaging for postharvest storage of mushrooms. A review. Fresh Produce, 1(1), 32-40.

Arvanitoyannis, I. (Ed.). (2012). Modified atmosphere and active packaging technologies. CRC Press.

Baur, E., Osswald, T. A., & Rudolph, N. (2019). Plastics handbook. Hanser Publications, Munich, 10, 9781569905609.

Bodbodak, S., & Moshfeghifar, M. (2016). Advances in modified atmosphere packaging of fruits and vegetables. In Eco-friendly technology for postharvest produce quality (pp. 127-183). Academic Press.

Czerwiński, K., Rydzkowski, T., Wróblewska-Krepsztul, J., & Thakur, V. K. (2021). Towards impact of modified atmosphere packaging (MAP) on shelf-life of polymer-film-packed food products: Challenges and sustainable developments. Coatings, 11(12), 1504.

Dobrucka, R., Cierpiszewski, R., & Korzeniowski, A. (2015). Intelligent food packaging-research and development. Log Forum, 11(1), 7-14.

Farber, J. N., Harris, L. J., Parish, M. E., Beuchat, L. R., Suslow, T. V., Gorney, J. R., Garrett, F. F.,&Busta, F. F. (2003). Microbiological safety of controlled and modified atmosphere packaging of fresh and fresh-cut produce. Comprehensive Reviews in Food Science and Food Safety, 2, 142-160.

Kargwal, R., Garg, M. K., Singh, V. K., Garg, R., & Kumar, N. (2020). Principles of modified atmosphere packaging for shelf life extension of fruits and vegetables: An overview of storage conditions. IJCS, 8(3), 2245-2252.

Kirkland, B. S., Clarke, R., & Paul, D. R. (2008). A versatile membrane system for bulk storage and shipping of produce in a modified atmosphere. Journal of Membrane Science, 324(1-2), 119-127.

Mangaraj, S., Goswami, T. K., & Mahajan, P. V. (2009). Applications of plastic films for modified atmosphere packaging of fruits and vegetables: a review. Food Engineering Reviews, 1, 133-158.

Marsh, K., &Bugusu, B. (2007). Food packaging—roles, materials, and environmental issues. Journal of Food Science, 72(3), R39-R55.

McMillin, K. W. (2008). Where is MAP going? A review and future potential of modified atmosphere packaging for meat. Meat Science, 80(1), 43-65.

McMillin, K. W. (2020). Modified atmosphere packaging. Food safety engineering, 693-718.

Mullan, M., & McDowell, D. (2003). Section A map gases, packaging materials, and equipment. Food Packaging Technology, 5, 303.

Mullan, M., & McDowell, D. (2011). Modified atmosphere packaging. Food and Beverage Packaging Technology, 263-294.

Nagarajarao, R. C. (2016). Recent advances in processing and packaging of fishery products: A review. Aquatic procedia, 7, 201-213.

Oliveira, M., Abadias, M., Usall, J., Torres, R., Teixidó, N., &Viñas, I. (2015). Application of modified atmosphere packaging as a safety approach to fresh-cut fruits and vegetables–A review. Trends in Food Science & Technology, 46(1), 13-26.

Oluwafemi, R. A., Edugbo, O. M., Solanke, E. O., &Akinyeye, A. J. (2013). Meat quality, nutrition security and public health: a review of beef processing practices in Nigeria. African Journal of Food Science and Technology, 4(5), 96-99.

Opara, L. U., Al-Jufaili, S. M., & Rahman, M. S. (2007). Postharvest handling and preservation of fresh fish and seafood. In Handbook of Food Preservation (pp. 169-190). CRC Press.

Opara, U. L. (2009). Quality management: an industrial approach to produce handling. Postharvest handling. A Systems Approach, 153-204.

Opara, U. L., Caleb, O. J., & Belay, Z. A. (2019). Modified atmosphere packaging for food preservation. In Food quality and shelf life (pp. 235-259). Academic Press.

Opara, U. L., Mditshwa, A. (2013). A review on the role of packaging in securing food system: Adding value to food products and reducing losses and waste. African Journal of Agricultural Research, 8(22), 2621-2630.

Parry, R. T. (Ed.). (2012). Principles and applications of modified atmosphere packaging of foods. Springer Science & Business Media.

Rao, C. G. (2015). Engineering for storage of fruits and vegetables: cold storage, controlled atmosphere storage, modified atmosphere storage. Academic Press.

Robertson, G. L. (2005). Food packaging: principles and practice. CRC Press.

Sandhya. (2010). Modified atmosphere packaging of fresh produce: Current status and future needs. LWT-Food Science and Technology, 43(3), 381-392.

Shayanfar, S., Kashaninejad, M., Khomeiri, M., EMAM, D. Z., &Mostofi, Y. (2011). Effect of MAP and different atmospheric conditions on the sensory attributes and shelf lifecharacteristics of fresh pistachio nuts.International Journal of Nuts and Related Sciences 2(23): 47–57.

Simpson, R., Carevic, E., & Rojas, S. (2007). Modelling a modified atmosphere packaging system for fresh scallops (Argopecten purpuratus). Packaging Technology and Science: An International Journal, 20(2), 87-97.

Sivertsvik, M., Jeksrud, W. K., &Rosnes, J. T. (2002). A review of modified atmosphere packaging of fish and fishery products–significance of microbial growth, activities and safety. International Journal of Food Science & Technology, 37(2), 107-127.

10

Edible and Biodegradable Packaging

Yogesh Kumar and Samandeep Kaur

Department of Food Engineering and Technology, Sant Longowal Institute of Engineering and Technology, Longowal, Punjab

10.1. Introduction

When it comes to packaging, plastics are still popular because of durability, versatility, economic production, and enhanced mechanical properties. However, with growing concerns about the impact of plastics-based packaging on the environment, edible and biodegradable packaging were introduced. The application of edible and biodegradable packaging promotes renewable resources, which greatly reduces environmental concerns and fulfil the packaging requirements. The sources employed to develop edible and biodegradable packaging are mainly plant-based biopolymers like cellulose, proteins, starch, and lipid. Moreover, the properties of the packaging material are decided by the proportions of these biopolymers. Hence, selecting a food commodity to be package is equally important as selecting suitable resources for the development of edible and biodegradable packaging.

Edible packaging consists of films, sheets, pouches and coatings, is defined as a thin layer of edible material formed on surface or wrapped on surface of foods with intention to be eaten with foods.For example, edible packaging is an integral component of pharmaceutical's capsules (soft-gel, hard-shelled) and coated tablets. The application of edible materials in food packaging is rapidly advancing with research and development in field of materials sciences. However, due to several limitations, which will discuss in further sections, edible packaging is an effective alternative of conventional packaging in some applications because of their unique ability to

- Improve shelf life and quality preservation
- Controls mass transmission
- Enhance sensory properties
- Act as physicochemical and mechanical barrier

- Carry and control the release of active ingredients
- Environment friendly and sustainable

The biodegradable packaging is intended to reduce the growing concerns of petroleum-based polymers (plastics), while serving the goal of packaging. Unlike plastics, biodegradable packaging is made of short chain molecules, which are eco-friendly, recyclable and easily decomposable by microbes within a short period of time. Biodegradable packaging offers flexibility to use in different form such as films to wrap the perishables and sealing containers, as coatings on fruits and vegetables to prevent microbial contamination and increase shelf life, andas bags to store food and cosmetics within metal containers. The above forms of the biodegradable packaging are mainly made of plant-derived materials, and involves tree cutting which can be regrown. However, other form like bioplastic is also getting popular now, which is made of polymers like cellulose, lactic acid, starch and protein from renewable sources.

Edible and biodegradable packaging materials made mostly of proteins or polysaccharides have sufficient mechanical and optical properties. However, they are extremely moisture-sensitive and have low water vapor barrier properties. This can be a downside when added to food items having considerable moisture levels, as the films can swell, melt or disintegrate upon contact with water. In such cases, lipid-based films are preferred as they do not interact with moisture; they show properties like opacity, relative rigidity, more susceptibility to oxidation, and poor mechanical properties. Considering these factors, native sources for the production of edible and biodegradable materials are incorporated with plasticizer and crosslinkers to conform with a range of basic technical criteria like features of the moisture and gas barrier, the solubility of water or lipids, color, and shape, as well as mechanical properties to maximize practicability and applicability. Hence, the successful production of high-performance edible and biodegradable packaging materials depends primarily on achieving three essential design requirements: better mechanical behavior, the preservation of adequate mechanical performance over the useful lifetime, and 100% achievement biodegradation in the soil after the end of the useful lifetime.

10.2. Edible Packaging Materials

10.2.1. Starch

Starch has been considered one of the most successful models for the production of edible and biodegradable packaging as it is odorless, tasteless, non-toxic, high abundance and comparatively economic. Starch is a biopolymer of glucose monomers joined in α 1,4 linkages and produced by most plants as energy storage. The native starch has two components; simple and linear which is called amylose, and branched polymer known as amylopectin. From applications point of view, amylose forms coherent, relatively strong and free-standing film due to linear and amorphous nature, whereas amylopectin-based films have found brittle and noncontinuous due to crystalline and non-linear nature. Therefore, the proportion of these two starch components decide the quality parameters of packaging film. However, the proportion of amylose in major starch sources like cereals grains (corn, wheat and rice) as well as tubers (potato, cassava and elephant foot yam) has been found significantly lower (10-30%) as compared to amylopectin. Hence, native starch is generally modified prior to film production to improve the mechanical as well as the barrier properties of film.

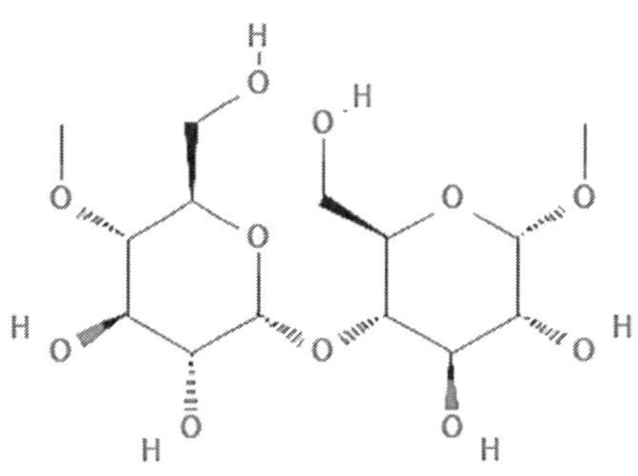

Amylose

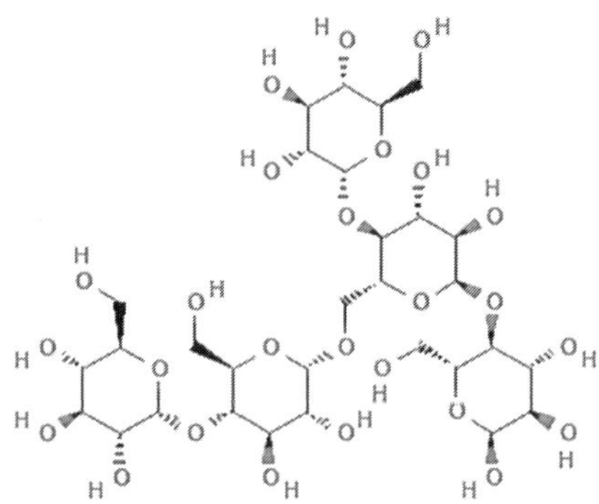

Amylopectin

Strengthening by natural fillers

Natural fibers (cellulose, Hemicellulose, Lignin, and Pectin) are of special interest in the biodegradable materials aspect due to their lower density and fiber-matrix adhesion. They have been utilized with natural polar matrixes, such as starch and protein to improve the mechanical properties. The tensile properties of thermoplastic wheat starch have been reported to have increased four times after reinforcement with cellulose fiber. Improved mechanical properties of starch-based materials have been used for various natural cellulosic fibers, such as sisal, cotton, bamboo, straw, and wood. Starch-based composite films have been found stronger as expected after the addition of rigid fillers like maize/wheat hulls, apricot and walnut shells, which are by-products of rigid particles from natural food supplies. By adding hulls with higher density and higher crystallinity, especially the corn hull, due to its flat structure, the gas barrier properties of the starch films have been found enhanced.

Blending with natural polymer

Polysaccharide and protein mixed films have shown unusual properties, such as a stronger gas barrier than any straight film. Previous studies have shown that gelatin and starch are immiscible, but different conditions, such as processing time, temperature, pH value, and solid concentration, influence their morphology and compatibility. The properties of starch-based materials can be greatly improved by mixing them with synthetic oil-based polymers.

10.2.2. Protein

Owing to the availability and biodegradability of animal and plant proteins like gelatin,collagen, fish myofibrillar protein, casein, gluten, corn zein, soy protein, and whey protein have been the most desirable sources of protein for packaging films. The properties of protein-based films depend on the source and concentration of the protein, the types of plasticizers and the concentration used, and other extrinsic influences. A complex series of chemical reactions are involved in creating the films; these are often affected by the experimental conditions.

Gluten

Based on the difference in solubility in aqueous alcohols, gluten proteins are mainly divided into gliadin (soluble) and glutenin (insoluble). The suitable film-forming property of gluten is based on its cohesiveness and elasticity. Two methods usually obtain gluten films: one is casting in a thin crust, and after casting, the drying process is conducted with aqueous alcoholics, and the other is boiling the protein solutions and then extracting the films produced on the surface of the solutions or thermos pressing. The features of cast films and thermos-pressed films are usually different: the former has higher elongation characteristics, and the latter has better tolerance to rupture. The stress-strain interaction between films obtained by the two techniques is also distinct, suggesting that the manufacturing process determines the protein network structure. The features of cast films and thermos-pressed films are usually different: the former has higher elongation characteristics, and the latter has better tolerance to rupture. The stress-strain relationship between films obtained by the two methods is also distinct, suggesting that proteins network structure influences the development mechanism.

Soy protein films

Soy protein isolate (SPI) films are based on soy protein isolate, which is a highly processed soy protein containing 90% protein and is made from soy flour by replacing the high mass of non-protein components. The increase in the molecular weight of soy proteins has shown different properties in the tensile fractions of different soy protein fractions. The boost in the molecular weight of increasing strength and elongation but not altering the water vapor barrier's properties. Soy protein films can also be produced through isoelectric precipitation from soy flour, soy milk carbohydrates, and fats.

Zein

Zein, a by-product of maize processing, accounts for approximately 45-50% of maize proteins. Owing to nonpolar amino acids in zein, zein proteins are soluble in alcohol but not in the water, which leads to the water vapor barrier properties of zein films. To prepare zein-based films, wet or solvent processes may be used, and polyols or fatty acids are usually selected as plasticizers. Firstly, zein, plasticizers, and some other agents are dissolved in a chosen solvent. Then, the film-forming solution is cast on a smooth and non-sticky surface. Finally, the solvent is evaporated, and the film can be peeled off from the surface by a free-standing film.

Casein

Casein, one of the milk proteins, consists of four major subunits: kappa-casein, beta-casein, alpha s1-casein, and alpha s2-casein, which make up 13%, 36%, 38%, and 10%, respectively. The unique properties of the four fractions of protein affect the ability of casein to form the film. Due to the strong interchain cohesion induced by its random-coil nature and many shaped intermolecular hydrogen, hydrophobic and electrostatic bonds, Casein can easily form films from aqueous solutions without further processing. Also, the presence of distributed polar amino acids along the casein chain makes the casein-based films a good barrier for non-polar molecules such as oxygen.

Whey

Whey, another milk protein, is a by-product of the processing of cheeses. The film-forming of whey-based films in aqueous solutions depends on the thermal denaturation of whey proteins. It is possible to change whey protein structure through heating; the three-dimensional structure of whey protein can be altered. It is possible to obtain plasticized whey protein-based films by heating the whey protein solution for a few minutes with a concentration of 8%-12% at 75-100 °C. The irreversible denaturation of whey proteins occurs in this way, and because of this, the films produced have a clear structure.

10.2.3. Cellulose and Hemicellulose

A linear polymer, cellulose, and a cross-linked polymer, hemicellulose, are two polysaccharides that contribute to the plant cell wall's structural properties. Cellulose is made up of glucose monomers arranged linearly, while hemicellulose comprises several polymers branching.

Owing to the high molecular weight of the glucose polymer, cellulose is a potential film-forming medium similar to other polysaccharides. It is neither meltable nor soluble in water or traditional organic solvents because cellulose has a strong crystalline structure and hydrogen bonds. Therefore, in its natural state, it does not form a gel/film. Therefore, to convert it into water-soluble products, such as cellulose derivatives, modifications of cellulose are also carried out. In nature, hemicelluloses are hydrophilic, and films made from these materials are typically hygroscopic, resulting in poor properties in high humidity environments.

10.2.4. Chitosan

Chitosan is a natural polymer, nontoxic, edible, and biodegradable, derived by chitin's deacetylation, which is the second most abundant biopolymer in nature after cellulose. In general, chitin is present in the exoskeleton of arthropods such as insects, crabs, shrimps, lobsters, and certain fungal cell walls. Besides that, against various microorganisms' classes, such as bacteria, fungi, and yeast, chitosan has antimicrobial activities. Chitosan has now been used to prolong food items' shelf life in edible coatings or films, such as fruits, meat, fish, and seafood. These researches explain the possibility of using chitosan as an antimicrobial packaging, which is a promising method of active food packaging. Unfortunately, due to its high moisture vulnerability, there are some drawbacks in adopting chitosan film for packaging.

10.2.5. Biocomposites

Edible biocomposites are multiphase packaging materials formed by blending natural fiber and biopolymers such as cellulose, starch, protein and chitosan. The synergetic interactions of different components not only improve the functional properties, but also help in overcoming the existing limitations. For example, starch and protein films have effective gas barrier properties; however, they are water sensitive and have good mechanical properties. On the other hand, lipids are effective against moisture migration due to their hydrophobicity, but have poor mechanical properties. Therefore, lipids are used with a hydrocolloids-based structural matrix to take the complementary advantages of both.

10.3. Biodegradable Materials

10.3.1. Polylactic Acid (PLA)

Polylactic acid (Lactic acid polymer) is a commercial biodegradable thermoplastic (easily recyclable) based on lactic acid ($C_3H_6O_3$). Lactic acid monomers are mainly produced by bacterial fermentation (70-90% of total lactic acid production) of carbohydrates-rich renewable resources like corn, sugar beets, and cane sugar. LA is either translucent or, at room temperature, resides in a yellowish glassy state. PLA is a linear polymer and lacks a versatile methylene segment in its backbone chain, resulting in the molecular chain's low mobility. There are multiple ester bonds in pristine PLA, which have low hydrophilicity and, therefore low biocompatibility. When PLA products are put in natural environments with high temperatures and high humidity, it is gradually hydrolyzed or biodegraded to form CO_2 and H_2O. By adding nucleating agents, which increase the crystalline and rigid amorphous fractions, the crystallinity of PLA can be increased. This can hinder PLA's chain mobility and allow it to with stand heat-induced distortions, resulting in improved heat resistance. A type of physical modification is fiber reinforcement. In the PLA matrix skeleton, the rigidity of fibers play a supporting role. The polymer's chain motion is reduced when PLA is heated, which improves the material's heat resistance. Using inorganic particles as the dispersed phase, PLA-based composites are usually prepared. During the preparation of the composites, the morphology of the dispersed process does not change. Inorganic particles usually have strong thermal properties, and the resulting composites of PLA have high resistance to heat.

10.3.2. Biopolyethylene

Bioplastics are produced from renewable biomass sources, such as vegetable fats and oils, corn starch, straw, woodchips, sawdust, and recycled food waste are biodegradable, environmentally friendly, or both. It is important to understand that bio-based plastics are not always biodegradable. The biodegradability of bioplastics depends on the raw material and the chemical structure, composition of the finished product, and the environment in which biodegradation of the product is required. The biodegradation of such materials is a mechanism involving microorganisms that are chemo-organotrophic prokaryotic (bacteria) and eukaryotic (fungi and protozoa), able to excrete enzymes (depolymerizes) that dissolve the polymer matrix and/or use water precursors and materials that store energy. Functionally, the biodegradation of plastic products must be disinfected by photochemical deterioration and non-specific biological attacks from the aging phase that usually occurs in the atmosphere. Such degradation can or may not be achieved by full bio-assimilation of the released compounds. Biodegradability is an end-of-life choice that allows the power of microorganisms present in the chosen environment to be harnessed on time to entirely eliminate biodegradable plastic products from the environmental compartment through the microbial food time chain in a timely and safe manner.

10.3.3. Nanocomposites

The developments in field of nanotechnology have opened many new possibilities to improve the functional and mechanical properties of packaging polymers. The nano-crystals and nano-fibers are used as fillers to reinforce the polymer matrix at the microscopic level. Due to the very large interaction between polymer matrix and filler, nanoparticles are more effective as compared to macroscopic reinforcement. However, the size of nano-crystals (10 and 400 nm) and nano-fibers (diameters: 2 to 20 nm and lengths 100 to thousands nm) depend on the origin and processing method. The nanocomposite is a multiphase solid material where one of the phases has one, two or three dimensions of less than 100 nanometers or structures having nano-scale repeat distances between the different phases that make up the material. Nanocomposites based on starch have continuously gained interest and have been widely produced and sold. Nano-crystals from natural origin (cellulose and starch) have been used to boost tensile strength as well as Young's modules. However, for coating and biodegradable films, metal-based nanoparticles (Titanium oxide, silver, gold, magnesium oxide, iron oxide and copper oxide) have been found to improve the shelf life of fruits. However, the safety issues of nanoparticles are major concerns as their possible migration to packaged food above certain limit can reduce the acceptability and organoleptic properties of food. In addition, the migration of nanoparticles can enter human lungs (70 nm), cells (50 nm) and blood (30 nm) to cause serious health problems.

10.4. Additives for Edible and Biodegradable Materials

10.4.1. Plasticizer

Plasticizers are colorless, odorless, low-molecular weight volatile compounds added to bio-polymer materials to modify films' functional properties by increasing their elasticity, flexibility, rigidity, and mechanical properties (Hanani et al., 2014a, 2014b). Food grade plasticizers can be manufactured from plants and agriculturally based glycerin derivatives. Polyols such as glycerol, mannitol, sorbitol, and xylitol; monosaccharides such as glucose, galactose, and sucrose; and fatty acids have been used as plasticizers in edible films. While studying the effect of plasticizer type and concentration on gelatin films, Bergo et al. (2013) observed that due to its higher hygroscopic nature, glycerol-containing films presented higher water absorption in comparison to other plasticizers (ethylene glycol, diethylene glycol and polypropylene glycol) studied. However, plasticizers may act as mechanical anti-plasticizers at concentrations less than 2.5% (Chang et al., 2006) and show significant effects on edible film's physical and textural properties (Cheng et al., 2002). However, most hydrophilic diluents such as polyols and saccharides have plasticizing effect rather than anti-plasticizing effects on the biopolymer-based film's physical and mechanical properties (Chang et al, 2006). Therefore, hydrophilic diluents are mostly used for the development of biodegradable films.

10.4.2. Emulsifier

Emulsifiers are additives added to mix two immiscible liquids. Commonly used emulsifiers in modern food production include mustard, soy and egg lecithin, glycerides, polysorbates, carrageenan, guar gum and canola oil. Due to high antioxidant and antimicrobial activity, essential oils such as citronella, cedarwood, are added during the emulsification process of edible films (Shen and Kamdem, 2015). Emulsifiers and active surface agents such as lecithin, mono- and di-glycerides or mono- and di-glyceride esters have often been added to emulsion film and coating formulations to improve their functional characteristics by stabilizing dispersed systems and increasing the particle distribution in composite emulsion-based edible films (Debeaufort and Voilley, 1995).

10.4.3. Hydrocolloids

Hydrocolloids are hydrophilic, high molecular weight polymers of plant, animal, microbial or synthetic origin. These functional compounds are used in food formulation for increasing food consistency, improving gelling effect, and controlling microstructure, texture, flavor, and shelf life. Some commonly used hydrocolloids are, alginate, carrageenan, carboxymethylcellulose, gum Arabic, pectin and xanthan gum (Williams and Phillips, 2000). All hydrocolloids are water soluble (completely/partially), and are mainly used as gelling agent or thickeners as they increase the viscosity of aqueous phase. They can also be used as emulsifier since their stabilizing effect on emulsions derives from an increase in viscosity of the aqueous phase of edible film. Edible films made from alginate exhibit good mechanical strength, and poor water resistance due to their hydrophilic nature (Skurtys et al., 2014).

10.4.4. Antimicrobials

Antimicrobial agents are substances added directly to food material or to packaging material, where they are released at time intervals to kill or halt the growth of different pathogenic microbes and maintain quality and shelf life of food materials. The antimicrobials are chosen as per the characteristics of target microorganism (i.e., Gram-negative/ gram-positive; aerobic/ anaerobic; spores/vegetative cells; acid/osmosis resistance; mesophilic/ thermophilic/ psychotropic). Antimicrobial agents can have microbial-cidal or microbial-static effects. For microbial static application, the agent should be able to maintain the desired concentration to prevent microbial growth throughout the storage period. Commonly used antimicrobial agent is Chitosan, which have good antimicrobial activity against many pathogenic and spoilage microorganisms, including gram-positive and gram-negative bacteria, molds and yeasts (Agullo et al., 2003; Leeet al., 2003). Antimicrobial activity of chitosan depends on deacetylation degree, pH value, molecular weight, surrounding matrix and type of microorganism (Dutta et al. 2009). Among volatile oils, garlic oil has been reported to have antimicrobial and antioxidant activity related with Sulphur compounds including allicin, diallyl disulphide and diallyl trisulfide (Nuutila et al., 2003; Pranoto et al., 2005). Other antimicrobial agents that are used in food industry are organic acids like benzoic acid, sorbates; enzymes like lysozyme, EDTA, nisin, glucose oxidase; bacteriocins like lauric acid; polymers like chitosan; natural extracts like grapeseed, clove, cinnamaldehyde.

10.4.5. Antioxidants

Foods such as meats, fish, whole milk powder, and sauces are prone to spoilage by lipidoxidation. To prevent this, the packaging industry has started the initiative of adding antioxidants to the food packaging which release antioxidants into the food material sustainably. On the basis of their mechanism of action, antioxidants can be classified as i) free-radical scavengers, ii) peroxides decomposers, iii) form oxidation complexes withmetals, iv) which act on singlet oxygen, v) which inhibit enzymatic activity crucial for auto-oxidation. Synthetic antioxidants such as butylated hydroxy toluene (BHT), butylatedhydroxy anisole (BHA) as well as, natural antioxidants such as tocopherols, catechins, quercetin, caffeic acid, and some essential oils from plants and herbs, have been incorporated into food packaging (LaCoste et al., 2005). The antioxidants can be incorporated into packaging materials by various means such as extrusion, coating on packaging/food item,antioxidant layer in multilayer packaging (Nerin, 2010). Complexation has been observed as an efficient tool for controlled-release of α-tocopherols in active packaging (Siro et al., 2006). Due to their biodegradable and generally recognized as safe (GRAS) nature, natural antioxidants are given preference over synthetic antioxidants while making active packaging. Antimicrobial, antioxidant as well as their mechanical properties of packaging materials has been significantly improved by the incorporation of antioxidant rich extracts of plants, fruits, grapefruit seeds, herbs such as oregano, rosemary, green tea etc. The antioxidant properties of these extracts are mainly attributed to the presence of bioactive compounds such as. quercetin, catechin, rosmarinic acid, carnosic acid, theaflavins, tocopherols, ascorbic acid, and various other phenolic and flavonoid compounds (Sanches-Silva et al., 2014).

10.4.6. Nanoparticles

Nanoparticles are small particles having size ranges between 1 to 100 nanometers. Despite having very small size, they exhibit significantly different physical and chemical properties when used in large materials. For example, mechanical and barrier properties of biopolymer based edible materials have been found to significantly improve with incorporation of nanoparticles. The improvement is mainly due to action of nanoparticles as a filler agent which reinforced the polymer matrices by filling void volumes and disrupting conformational positions. Moreover, they also immobilize those adjacent molecules groups and orientations of the polymers. For food-based applications, natural source-based nanoparticles like starch nanoparticles, cellulose nanofibers, and chitosan nanoparticles are best option due to their nontoxic and neutral nature with significant effect on physical-mechanical performances of films.

10.5. Methods for Edible Film Formation

The edible films can be formed by simple coacervation, complex coacervation orthermal gelation/coagulation mechanism. Coacervation is the easiest method, which involves only one polymer in an aqueous or aqueous/organic solution. In contrast, complex coacervation includes two oppositely charged polymers in the solution. In both methods, coacervation separates the polymeric solution into two immiscible liquid phases: one with a dense, cohesive and structural matrix known as coacervate phase, and other with weak interacting forces known as dilute equilibrium. Therefore, the functional and structural performance of edible films largely depend on the:

- Interaction forces such as covalent bonds, ionic bonds and H-bonds
- Polymeric structure like molecular weight, branching and polarity
- Film forming conditions like solvent types, pH, temperature, pressure and ageing time
- Content of plasticizers and additives
- Film formation procedures such as casting and extrusion

Thermal gelation is used for those polymers, which are unable to interact with solvent at lower temperature. For example, starch is very poorly soluble in cold water, but when suspension is heated, its granules start swelling and form a well-structured gel. Hence, this method is very common in formation of starch and protein films.

For edible coatings of fats, waxes or resins, thermal processes like melting, andsolidification are used. However, since these biopolymers are easily soluble in organic solvents like ethanol, toluene, hexane, etc. solubilizing them followed by evaporation of the solvent can also be used.

10.6. Properties of Edible and Biodegradable Packaging Materials

10.6.1. Mechanical Properties

The ability of any film to sustain external stress depends on its mechanical strength and flexibility, adequate amount of these properties is important to maintain integrity and barrier properties of the packaging film. The mechanical properties like tensile strength, Young's modulus and percent of elongation at break (EAB) of native biobased edible packaging are not satisfactory. For starch-based film, tensile strength has been found to vary between 24-80 MPa, which is surprisingly higher than tensile strength of LDPE films (8-20 MPa). For PLA, tensile strength has been reported to vary between 40-63 MP, but have EAB very less 9% which suggest brittle nature of film. However, addition of plasticizers (≈5%glycerol) significantly improved the tensile strength and EAB of gelatin and starch films. In addition, the tensile strength, elongation at break and puncture test values of gelatin-polyols/sucrose/lipid plasticized film has been found in range of 0.99 to 38.79 MPa, 1.25–294.5% and 2.44–54.8 N, respectively. Owing to its hygroscopic nature, glycerol increases the mobility of polymer chains, and hence leads to higher plasticization effect. Addition of different oils have also been found effective in improving the mechanical properties of edible films. The EAB of films have been found significantly increased after adding 10% concentration of essential oils while the tensile strength of chitosan films slightly compromised from 33 MPa to 29.4 MPa when oil content increased from 10% to 20%. The mechanical properties of hydroxypropyl methylcellulose edible films were improved significantly by addition of chitosan/tripolyphosphate nanoparticles.

10.6.2. Barrier Properties

The permeability of gases (such as O_2, CO_2, and N_2), water vapor, aroma compounds and light through any packaging barriers have deteriorating effect on the quality of packaged foods. Due to hydrophilic nature of biobased edible packing materials, they have poor barrier properties against polar substance like water vapor. It has been found that water vapor transmission through starch-based film is four to six times higher as compared to synthetic polymers made from petroleum-based polymers. Therefore, their application in high humid environment is avoided. However, biobased edible packaging has shown comparatively better barrier properties against gases (O_2, CO_2, and N_2) and against nonpolar substances like aromas, and oils.

10.7. Applications

Biodegradable films are prepared from edible material that act as a barrier to external elements (factors such as moisture, oils, gases and vapors) and thus protect the product, extend its shelf life and improve its quality (Suyatma et al., 2005). Different food ingredients, derived from meats, cereals, nuts, fruits and vegetables, are being used to produce edible/biodegradable films. These films act as novel packaging systems for their biodegradable nature and ability control the release of active compounds such as antimicrobial agents, and antioxidants, for food preservation and protection (Tharanathan, 2003). Application of edible films to coat fruits and vegetables is an appealing way to eliminate plastics and at the same time films with active ingredients act as insecticide or insect repellent (Kyrikou and Briassoulis, 2007). Edible films are helpful in delaying the moisture content of food and the environment, controlling the amount of food breathing by reducing the amount of oxygen absorbed and selective blocking of CO_2 and vapor, reducing the migration of fats, maintaining the structure of food, including food additives, preventing the migration of aroma and flavoring and colored food products to the environment and between heterogeneous food components, preventing microbial corruption during long-term storage, increasing the nutritional value of the product, protecting the product against mechanical and physical damage, reducing the amount of packaging material and complexity of packaging. These films in food industry can control or prevent the occurrence of reactions occurring within the package. Biodegradable packaging according to edible films have more importance due to natural ingredients and lower environmental pollution. In this regard, the ability of such films as cover of antimicrobial and antioxidant factors and other active factors to enhance shelf-life, increase quality, monitor pathogens and increase organoleptic features of the materials have different use for them in food packaging industry (Nafchi et al., 2017). The use of carrageenan as edible films and coatings already covers various fields of the food industry such as application on fresh and frozen meat, poultry and fish to prevent superficial dehydration, sausage-casings, dry solids foods, oily foods, etc. (Skurtys et al., 2014).

References

Ali, A., Yu, L., Liu, H., Khalid, S., Meng, L., & Chen, L. (2017). Preparation and characterization of starch-based composite films reinforced by corn and wheat hulls. Journal of Applied Polymer Science, 134(32), 45159.

Alves, J. S., Dos Reis, K. C., Menezes, E. G. T., Pereira, F. V., & Pereira, J. (2015). Effect of cellulose nanocrystals and gelatin in corn starch plasticized films. Carbohydrate Polymers, 115, 215-222.

Atarés, L., & Chiralt, A. (2016). Essential oils as additives in biodegradable films and coatings for active food packaging. Trends in Food Science & Technology, 48, 51-62.

Beck, M. I., Tomka, I., & Waysek, E. (1996). Physico-chemical characterization of zein as a film coating polymer: A direct comparison with ethyl cellulose. International Journal of Pharmaceutics, 141(1-2), 137-150.

Byun, Y., Rodriguez, K., Han, J. H., & Kim, Y. T. (2015). Improved thermal stability of polylactic acid (PLA) composite film via PLA–β-cyclodextrin-inclusion complex systems. International Journal of Biological Macromolecules, 81, 591-598.

Chen, Y., Yu, L., Ge, X., Liu, H., Ali, A., Wang, Y., & Chen, L. (2019). Preparation and characterization of edible starch film reinforced by laver. International Journal of Biological Macromolecules, 129, 944-951.

El Miri, N., Abdelouahdi, K., Barakat, A., Zahouily, M., Fihri, A., Solhy, A., & El Achaby, M. (2015). Bio-nanocomposite films reinforced with cellulose nanocrystals: Rheology of film-forming solutions, transparency, water vapor barrier and tensile properties of films. Carbohydrate Polymers, 129, 156-167.

Fredon, E., Granet, R., Zerrouki, R., Krausz, P., Saulnier, L., Thibault, J. F., & Petit, C. (2002). Hydrophobic films from maize bran hemicelluloses. Carbohydrate Polymers, 49(1), 1-12.

Hansen, N. M., & Plackett, D. (2008). Sustainable films and coatings from hemicelluloses: a review. Biomacromolecules, 9(6), 1493-1505.

Jauregi, P., &Welderufael, F. T. (2010). Added-value protein products from whey. Nutrafoods, 9(4), 13-23.

Jiang, T., Duan, Q., Zhu, J., Liu, H., & Yu, L. (2020). Starch-based biodegradable materials: Challenges and opportunities. Advanced Industrial and Engineering Polymer Research, 3(1), 8-18.

Jin, F. L., Zhang, H., Yao, S. S., & Park, S. J. (2018). Effect of surface modification on impact strength and flexural strength of poly (lactic acid)/silicon carbide nanocomposites. Macromolecular Research, 26(3), 211-214.

LaCoste, A., Schaich, K. M., Zumbrunnen, D., & Yam, K. L. (2005). Advancing controlled release packaging through smart blending. Packaging Technology and Science: An International Journal, 18(2), 77-87.

Li, X., Qiu, C., Ji, N., Sun, C., Xiong, L., & Sun, Q. (2015). Mechanical, barrier and morphological properties of starch nanocrystals-reinforced pea starch films. Carbohydrate Polymers, 121, 155-162.

McHugh, T. H., & Krochta, J. M. (1994). Water vapor permeability properties of edible whey protein-lipid emulsion films. Journal of the American Oil Chemists' Society, 71(3), 307-312.

Nerín, C. (2010). Antioxidant active food packaging and antioxidant edible films. Oxidation in foods and beverages and antioxidant applications. In. Volume 2: Management in Different Industry Sectors, 496-515.

Panaitescu, D. M., Frone, A. N., Ghiurea, M., & Chiulan, I. (2015). Influence of storage conditions on starch/PVA films containing cellulose nanofibers. Industrial Crops and Products, 70, 170-177.

Peelman, N., Ragaert, P., Ragaert, K., De Meulenaer, B., Devlieghere, F., & Cardon, L. (2015). Heat resistance of new biobased polymeric materials, focusing on starch, cellulose, PLA, and PHA. Journal of Applied Polymer Science, 132(48).

Qian, S., Zhang, H., Yao, W., & Sheng, K. (2018). Effects of bamboo cellulose nanowhisker content on the morphology, crystallization, mechanical, and thermal properties of PLA matrix biocomposites. Composites Part B: Engineering, 133, 203-209.

Sanches-Silva, A., Costa, D., Albuquerque, T. G., Buonocore, G. G., Ramos, F., Castilho, M. C., & Costa, H. S. (2014). Trends in the use of natural antioxidants in active food packaging: a review. Food Additives & Contaminants: Part A, 31(3), 374-395.

Shimada, K., &Cheftel, J. C. (1989). Sulfhydryl group/disulfide bond interchange reactions during heat-induced gelation of whey protein isolate. Journal of Agricultural and Food Chemistry, 37(1), 161-168.

Shukla, R., & Cheryan, M. (2001). Zein: the industrial protein from corn. Industrial Crops and Products, 13(3), 171-192.

Siró, I., Fenyvesi, É., Szente, L., De Meulenaer, B., Devlieghere, F., Orgoványi, J., & Barta, J. (2006). Release of alpha-tocopherol from antioxidative low-density polyethylene film into fatty food simulant: Influence of complexation in beta-cyclodextrin. Food Additives and Contaminants, 23(8), 845-853.

Soliman, E. A., MohyEldin, M. S., & Furuta, M. (2009). Biodegradable zein-based films: influence of γ-irradiation on structural and functional properties. Journal of Agricultural and Food Chemistry, 57(6), 2529-2535.

Soykeabkaew, N., Laosat, N., Ngaokla, A., Yodsuwan, N., & Tunkasiri, T. (2012). Reinforcing potential of micro-and nano-sized fibers in the starch-based biocomposites. Composites Science and Technology, 72(7), 845-852.

Tharanathan, R. N. (2003). Biodegradable films and composite coatings: past, present and future. Trends in Food Science & Technology, 14(3), 71-78.

11

Packaging of Food Products

Anbu Raj[1], Anand Kishore[1] and Khushbu Kumari[2]

[1]*National Institute of Food Technology and Entrepreneurship and Management Kundli, Haryana, India*
[2]*National Dairy Research Institute, Karnal, Haryana, India*

11.1. Introduction

Food packaging is the integral part of the food industry. Packaging may be defined as the enclosure of products, items or packages in a wrapped pouch, bag, box, cup, tray, can, tube, bottle or other container form to perform one or more of the following functions: containment, protection, preservation, communication, utility and performance. If the device or container performs one or more of these functions, it is considered a package. The major roles of the packaging of food are the containment and preservation of foods by protecting the food from physical, chemical and microbial spoilage. Besides, packaging performs some other functions like communication which means giving the information about the product, transportation of the food production and so on. Food products are classified as perishable, semi- perishable and non-perishable food based on the spoilage. Selecting appropriate packaging material and packaging technique for various food products like meat and poultry, fruits and vegetables, milk and milk products, bakery and confectionery and beverages is essential to maintain the quality of food product. Selection of the ideal packaging material for specific food requires prior knowledge about the packaging materials and its properties, type of the food products and specific spoilage mechanism of the particular food. Above all, from economical point of view, packaging material cost and capital for the establishment of the packaging technique also plays crucial role in selection of a particular packaging material or technique. The important consideration required for selection of suitable packaging system is shown in table 1. This chapter will give you the brief idea about the type of packaging materials and type of the technique and recent trends involved in the packaging of various food products.

Table 1: Parameters for selection of effective packaging

Parameter	**Considerations**
Weight	• The strength of the packaging box • An appropriate box should be able to hold the weight of the content being packaged • Weight of food product should be as per regulation
Size and Shape	• Size and shape of product is important to design the suitable packaging system • Odd-shaped or rounded packaging may need extra attention
Form of Contents	• Special packaging materials and packing techniques are required for packaging of solid, semi-solid, liquids and powders
Value of Contents	• Extra cushioning and protection may be required for high value goods
Fragility	• Extra cushioning and protection are required for fragile items • Apply special handling label
Regulations	• Follow the standard regulation for packaging of particular product • Some regulated items may require specialist packaging

11.2. Packaging of Horticulture Products

Consumer demands the minimally processed and fresh fruits and vegetable. This will be the challenge for the food technologist to maintain the freshness of packaged horticultural products. Fruits and vegetables are more prone to spoilage so that most of the fruits are classified under the perishable food. Exceptionally some of the fruits and vegetable comes under the semi perishable food like onion and potato. Spoilage factors of these commodities are mainly respiration rate, ethylene production rate, microbial growth and enzymatic reactions in the case of cut fresh fruits and vegetable.

11.2.1. Packaging of Fresh Fruit and Vegetables

Fresh fruits and vegetables are more susceptible to physical damage during transportation and storage. Physical damage causes the bruising on the surface of these fresh fruits and vegetable which reduce the product palatability. These fresh commodities undergo the biological process like ripening and ageing involves the breakdown of plant tissue. Moreover, moisture loss is possible during the storage which causes the shrinkage of the fresh produce. The packaging of fruits and vegetables mainly focuses on eradicating the previously mentioned major spoilage factors. However, the improper packaging systems of these commodities will accelerate the spoilage.

Fruits and vegetables are classified based on perishability as follows.

- Soft fruits(berries and grapes etc.) are more perishable,

- Hard fruits (apple and citrus fruits etc.) are less perishable, resist to damage and has less respiration rates,
- Root vegetables (Potato, Onion etc.) are less perishable
- Stem vegetables (celery, asparagus etc.) - high perishable and prone to moisture loss.

Packaging material selection is based on the nature of spoilage and it perishability. Conventionally, fresh fruits and vegetables are packed in corrugated fibreboard carton and wooden boxes to minimise the physical damage. Nowadays, these are replaced by plastic crates formed by the injection molding of HDPE and Polypropylene as they are low in weight when compared to the wooden boxes. Also Tray, cups, pads and tissue paper wraps are used as primary internal packaging materials for the delicate fruits as extra protection for the damage. However, considering its perishability, these horticultural commodities are mostly sold unpacked in retail or wholesale using polyethylene (PE) bags or net. These PE bags are perforated which allows the commodities to breathe. Semi-rigid containers covered with the cellophane or polystyrene or cellulose acetate are used to pack the soft fruits to minimise the mechanical damage. Hard fruits are packed in the open tray or plastic film warps. Stem vegetables are packed in the cellophane or polyethylene with ventilation because of their moisture barrier properties. Root vegetables like Potatoes, Radishes, Carrots, beets should be pre-packed for the distribution and market sale. Perforated Polyethylene and polypropylene films are used for this purpose. Additionally, wax coating is commonly applied for some fruits like citrus fruits, apples, tomatoes. This waxing process reduces the moisture loss by controlling the transpiration of the commodities and extends the shelf life.

As living things, fruits and vegetable continue to respire after harvesting. Generally, respiration rate is inversely proportional the shelf life of fruits and vegetables respiration rate of fruits and vegetables. MAP technique used for the packaging of the fruits and vegetable to increase the shelf life of the product by controlling the respiration rate. The basic principle of MAP is modifying the composition of gases like nitrogen, oxygen, and carbon-di-oxide in the headspace of the packaging or the container. Gas composition of MAP for packaging headspace depends on the type of the fruits and vegetables need to be packed. Gas composition and storage temperature for the different types of fruits and vegetable and its effect on shelf life are studied by the various researchers and tabulated (Table 2).

Table 2: Optimum gas composition for the MAP of different horticultural produces (Soltani et al., 2015)

	Products	**Temperature (^0C)**	**Recommended modified atmosphere**		**Potential**
			O_2(%)	**CO_2(%)**	
Fruits					
	Apple	0-5	1-3	1-5	Excellent
	Apricot	0-5	2-3	2-3	Fair
	Avocado	5-13	2-5	3-10	Good
	Banana	12-15	2-5	2-5	Excellent
	Cherry, sweet	0-5	3-10	10-15	Good
	Fig	0-5	5-10	15-20	Good
	Grape	0-5	2-5	1-3	Fair
	Grapefruit	10-15	3-10	5-10	Fair
	Kiwi fruit	0-5	1-2	3-5	Excellent
	Lemon and lime	10-15	5-10	0-10	Good
	Mango	10-15	3-5	5-10	Fair
	Olive	5-10	2-3	0-1	Fair
	Orange	5-10	5-10	0-5	Fair
	Peach and nectarine	0-5	1-2	3-5	Good
	Pear	0-5	1-3	0-3	Excellent
	Persimmon	0-5	3-5	5-8	Good
	Plum	0-5	1-2	0-5	Good
	Raspberry and other berries	0-5	5-10	15-20	Excellent
	Red currant	0-2	5-10	15-20	Excellent
	Strawberry	0-5	5-10	15-20	Excellent
Vegetables					
	Asparagus	0-5	Air	5-10	Excellent
	Bean	5-10	2-3	4-7	Fair
	Broccoli	0-5	1-2	5-10	Excellent
	Brussels sprout	0-1	2-4	4-6	Excellent
	Cabbage	0-5	2-3	3-6	Excellent
	Cantaloupe	3-7	3-5	10-15	Good

	Products	Temperature (^0C)	Recommended modified atmosphere		Potential
			O_2 (%)	CO_2 (%)	
	Celery	3-5	1-4	0-5	Good
	Chicory	5	2-3	5-10	Excellent
	Corn (sweet)	0-5	2-4	5-10	Good
	Cucumber	8-12	3-5	0	Fair
	Leek	0-5	1-2	3-5	Good
	Lettuce	0-5	1-3	0	Good
	Onion	0-5	1-2	2-5%	Good
	Pepper	8-12	3-5	0	Fair
	Spinach	0-5	Air	0-20	Good
	Tomato	12-20	3-5	0-3	Good

Active and Passive MAP is generally used for packaging of fruits and vegetables. In active MAP, atmosphere is modified actively by replacing the gases in the headspace of the package. The advantage of this active MAP is the quick establishment of EMAP (Equilibrium modified atmosphere packaging). Additionally, active substances also used to extend the shelf life of fruits and vegetable which includes ethylene scavengers, oxygen scavenger and carbon-di-oxide emitters inside the suitable packing system. CO_2 emitters and O_2 scavengers are used to control the respiration rate. Ethylene scavengers are used to control the ripening of the fruits which increases the longevity of the fruits.

In passive MAP, final equilibrium in the packaging atmosphere is achieved naturally by the respiration of the commodities. Passive MAP is attained by matching the respiration rate of the commodities with polymeric film characteristics. Moreover, Permeability and respiration rate also depends on the temperature. So, Temperature and permeability of the film are selection criteria of polymeric film for passive MAP. The design of the passive MAP requires the knowledge of the parameters like products respiration rate, film permeability, optimum CO_2 and O_2 concentration for the particular type of the product.

11.2.2. Packaging of the Canned Fruits and Vegetables

Minimally processed fruits and vegetable are canned for value addition and facilitating the international export of fruits and vegetables. The canning process let the people to enjoy their favourite fruits and vegetable in off season. Generally spinach, pineapple, carrots, beans (cut and whole), beets, apples, corn, peas, tomatoes, peaches, pears, apricots, and cranberries are canned. Canning is also done for fruit juice, ketchup and other value added products from fruits and vegetable. Canning process is the type preservation technique which involves the decontamination of packaging material as well as the food products by the application of heat and preventing recontamination by hermetic sealing. Often, fruits and vegetables are cut into pieces, blanched and they are kept in sugar syrup or brine solution in cans. As the term implies, metal cans are used as the packaging material for the canning. Generally two and three pieces can are used for canning. Low carbon steel and tinplate cans are used to make the cans for canning. Tinplates are lacquered with lacquering agents like oleoresins, phenolic and epoxy phenolic, vinyl and acrylic lacquers. Lacquering process is given to tin cans to avoid the leaching of the tin into the food product upon long storage. Advantages of using metal cans in the canning include their rigidity and suitability for heat sterilization process. Also they have very low gas and water vapour transmission rate thus provides the hermetic sealing of the container. Also, metal cans block the ambient light, so that spoilage reaction accelerated by light is prevented. Thermophillic anaerobic spoilage of the canned fruits and vegetable can be identified by the bulging of the cans due to the release of large amount H_2 and CO_2 gases.

Another similar term to canning is bottling also called as home canning. This is the widely used home preservation technique of fruits and vegetables. Different types of glass containers like bottles, jars, tumblers, jugs and vials are used in bottling instead of metal cans. These glass containers are sterilized in pressure cooker. Advantage includes glass containers do not develop the metallic flavours as they are chemically inert, they will look attractive due to their transparent nature and their reusability. Metallic lids with gasket on the rim are used for air tight sealing.

11.2.3. Packaging of the Dehydrated Fruits and Vegetables

Dehydration is employed to fruits and vegetables as a preservation technique. These dehydrated fruits and vegetable need to be protected from the moisture, dust, and from insects and rodents. As the spoilage factors for dehydrated food products is comparatively less than fresh fruits and vegetable, cost of the packaging is less. This is also one of the reasons for the dehydration of the fruits and vegetables. These dried fruits and vegetables should be cooled before the packaging or else sweating (condensation of water vapour) will take place and spoil the product. Conventionally, these dehydrated fruits and vegetables are packed in tin cans, bins and drums and glass jars with metal twist off caps depend on the type of the product. These dried products have low moisture content so that packaging material have to maintain this characteristic of the dried product which means packaging material should be low permeable to water vapour. Polypropylene trays with thermo sealed using OPA/PE, PA/PP or PA/PE laminates is used for the packaging of the dried fruits like raisins and apricot. Vacuum packaging with tight seal using polypropylene film with thickness of 25 mm is used for the dried vegetables. MAP with higher amount of N_2 and less O_2 is also used for inhibition of aerobic microbes for the packaging of the dried fruits and vegetables. Corrugated fibre board boxes with moisture absorbers like silica gel is used for the secondary packaging for dried fruits and vegetables.

11.2.4. Packaging of Frozen Fruits and Vegetable

Freezing is the simplest preservation technique for fruits and vegetable. Freezing inactivates the enzymes and microbes which increase the shelf life of the fruits and vegetable. Size reduction and blanching are the pre-treatments employed prior to the freezing. Frozen horticultural commodities should be properly packed to maintain its quality till the end consumer. The packaging used for the packaging of frozen fruits and vegetables should moisture vapour proof. Moisture vapour proof is the ability packaging material to prevent the evaporation of the frozen horticultural commodities. Rigid plastic containers and glass containers come under moisture vapour proof packaging materials. However non rigid packaging materials made up of moisture vapour resistant also can be used. These frozen commodities should be kept away to avoid the oxidative deterioration. Another factor for the consideration is durability that the packaging material should withstand low temperature (PE can withstand -40^{0}C) since some of the polymeric material become brittle when exposed to lower temperature for the longer period of time. Materials like HDPE, Rigid PET, glass jars,tin cans are used for frozen fruits and vegetables as they have low permeability to vapour and oxygen. Along with that MAP like vacuum packaging and gas flushing are used to eliminate the oxygen inside the package which significantly increases the longevity.

11.3. Packaging of the Milk and Milk Products

11.3.1. Packaging of the Fluid Milk

Liquid milk is usually collected and stored as the bulk in tanks and drums made up of galvanized iron or steel tin. Aluminium can are also used for the short term storage.Milk has a complex physico-chemical properties, It consists of a water solution of salts, lactose, proteins in a colloidal form and milk-fat solids are present in a partially emulsified suspension. The fluid milk is differentiated by its process and composiotion so the selection of suitable packaging material and techniques may be different for the different type of milk. In general the indices of failure for milk are: leakages from milk pouch, Milk is prone to oxidation of fat, light, heat affects the quality and loss of vitamins. The packaging requiremnet for different type of milk is shown in Table 2.

Table 2: Packaging requirement of different type of milk

Classification	**Products**	**Significance**	**Packaging Requirement**
Liquid Milk Products	Toned milk, Double toned milk, Full cream milk, Flavoured milks, etc.	Perishable due to high moisture, ideal for microbes and presence of light sensitive components	Protection against light and oxygen
Dried Milk Products	Milk powder (skim and whole), Whey protein concentrate/isolate, etc.	Low moisture, high fat and light sensitive	Protection against water vapour, light and oxygen
Fat rich products	Butter, spreads, etc.	High fat with fat soluble vitamins and flavour	Protection against water vapour, light and oxygen and aroma protection
Traditional Indian Dairy Products	Burfi, peda, paneer, gulabjamun, rasogolla, ghee, etc.	High fat, low moisture products with characteristic flavours,	Protection against moisture, light, preservation of aroma
Long-life products	UHT milk, Sterilized milk, retort processed products (like paneer curry), canned products (rasogolla, gulabjamun, condensed milks), etc	Sterile, medium to high fat, characteristic flavours with minimum head space in the pack	Protection against microorganisms, light, oxygen, moisture and container surface and preservation of aroma

11.3.2. Packaging of the Pasteurized Milk

Milk pasteurization involves the killing of the selective vegetative microorganism in the raw milk. LTLT and HTST (Flash Pasteurization) provides the short time preservation of milk. These two pasteurization process facilitates lesser reduction in microbial load and nutrition quality. In pasteurization milk is cooled rapidly after the heat treatment to cease the microbial growth. After cooling process the milk is subjected to the packaging. Packaging of pasteurized milk has been done in aseptic condition to prevent the recontamination of packaging milk. Aseptic packaging of the milk refers to filling the pasteurized milk in sterile container or packaging material in the aseptic condition (germ-free environment). For this purpose, the packaging material used for the packaging of the milk should be resistant to UV treatment that is employed for the sterilization of the packaging film. The other preferred characteristics of packaging materials for packaging of milk include oxygen barrier properties, odour barrier properties and other mechanical properties. Conventionally, Pasteurized milk packed in glass bottles or LDPE lined cartons. These are replaced by the LDPE or LLDPE films and are currently used in the markets as cost effective alternative to glass bottles for the packaging of the pasteurized milk.

11.3.3. Packaging of the UHT Milk

In contrast to the LTLT and HTST pasteurization, UHT pasteurized milk can be stored for longer time but UHT causes high nutritional degradation. Ultra-high temperature is the heat treatment upto 135^{0}C given to the milk to kill all the vegetative cells and spores. However, Not only the heat treatment gives the longer shelf life of the milk, the proper packaging of the heat treated milk also necessary to extend the shelf life of the UHT milk. It uses the aseptic packaging just like the packaging of pasteurized milk. These containers (tetrapak) are made up of lamination of paper board/Al foil/PE. In this multi-laminated pouch aluminium layer provides the higher barrier property. These containers maintain the quality attributes of milk. These UHT treated can be stored up to 3 to 6 months without refrigeration.PET bottles also provides extended shelf life for the UHT treated milk.

11.4. Packaging of Butter

Butter is comprised of the 80% of fat and 15 % of moisture and is more susceptible to the moisture loss, oxidative rancidity and mould growth.. Hence the packaging material should have the low oxygen transmission and water vapour transmission rate, also the packaging material of the butter should preserve its flavour from the butter and prevent flavour absorption from outside environment. Therefore, the packaging material should be less permeable to the volatile of the butter. Generally, butter block are wrapped with parchment paper / polyethylene as primary packaging and paperboard box extrusion coated with PP as secondary packaging. Flexible packaging materials made by vegetable parchment paper, Al foil, paper board carton is used for packaging of the butter. Parchment paper provides wet strength and al foil act as the barrier for the oxygen. PP tubs or pots made by the injection moulding can also be used for packaging of butter. HIPS (High Impact polystyrene) extruded with PE or PP or PET or PVdC and EVOH are also used and recently, al-foil trays with heat sealable PVdC- Cellophane is being used.

11.5. Packaging of the Cheese

Cheese is widely used dairy products which contains more fat and moisture. Major cheese spoilage and quality loss occurs in terms of moisture loss, oxidative reaction, alteration in pH, change in water activity and microbial spoilage. Packaging material should have oil or grease resistance and better barrier properties for the packaging of cheese. Cheese is usually packed under vacuum atmosphere to eliminate the oxygen to prevent the cheese from oxidation.. The coextruded polymer used in the cheese packaging is PVdC/ Polyester. Generally, cellophane, coextruded PVdC / Polyester PP/EVA/ LDPE laminate is used for the packaging of the cheese. Cottage cheese is packed in the polystyrene tubes in vacuum condition. Plastic lined metal cans also be used for the packaging of the cheese. Nylon coated with LDPE on both side can also be used.protection against light and oxygen is necessary to improve the qulaity of cheese, hence MAP could extend the shelf life of cheese.

11.6. Packaging of the Milk Powders

Milk powders are made from fluid milk by drying process. Milk is classified as skim milk powder and whole milk powder based on composition of milk used for the production of the milk powder. Milk powders are hygroscopic in the nature and during the storage, the milk powder majorly undergoes the oxidative rancity and non-enzymatic browning reaction due to reaction between the lysine and lactose which is present in the milk powder ...Packaging material for the packaging of milk powder should have low barrier properties and ensure the blocking of light. Whole milk powder is susceptible to the oxidative rancidity as it contains fat in it. These are shelf-life limiting factors for the milk powders should be taken into consideration while choosing the packaging material for milk powder.

Skim milk powders are generally packed in the PE and LDPE. Mostly, single layer high water vapour barrier film is enough to protect the skim milk powder from spoilage because it contains very less or negligible fat content. However, whole milk powder should be packed no oxygen or low oxygen condition to control the oxidative reaction because it contains high fat conatent. Laminates like Paper Board/ PE/ Al foil/PE or LDPE/Al foil/PET can be used for the packaging of the whole milk powder. It is mentioned that aluminium foil layer in the laminate enhance barrier properties of the laminates. Hermetically sealed metal cans are also used for the longer shelf life of the milk as metal cans has more barrier properties. MAP of milk powder uses 100% N_2 for the increased shelf life of milk powder. Sometime MAP and vacuum packaging of milk powder also employed for the complete elimination of oxygen.

11.7. Packaging of Cereals, Snack Foods and Confectionery

11.7.1. Packaging of Cereals and Pulses

Wheat, paddy, barley, maize and millet like pearl millet, finger millet etc. comes under the cereals. Gram, dhal, beans are classified as pulses. These cereals and pulses are classified into wide range of products includes whole grains, milled grains, breakfast cereals, split pulses, and ready to cook mixes. These pulses and cereals products can undergo many spoilage modes. Factors like respiration, oxygen availability, humidity, temperature, spillage losses, microbial spoilage and other biological caused by rodents and insects. These factors should be taken into account for the packaging material selection.

Conventionally, whole cereal and pulses are packed in the jute bags for bulk packaging. This jute has several disadvantages like less resistance to UV radiation and corrosive chemicals , less availability and cost. HDPE or PP woven sack are used as cost effective alternative for jute bags. For consumer level packaging, LDPE or LLDPE films, PP or BOPP films, laminates like BOPP/LDPE, CPP (Cast Polypropylene)/LDPE or Polyester/LDPE are used. These laminates are used for premium products where retention of flavour is necessary.

However, Suji, Maida, Atta tends to absorb which cause the hydrolytic rancidity. These milled products starts to deteriorate when moisture content exceeds 13%. Along with that these milled products should be protected from microbial spoilage and insect attack. Traditionally, gunny bags are used for bulk packaging of the milled products. This gunny bags are more susceptible to the microbial contamination and infestation. HDPE or PP woven sacks are used nowadays. These sacks are laminated with LDPE to provide additional moisture barrier. For consumer level packaging of milled products LDPE films, LLDPE films, HMHDPE films, PP films, BOPP films, Co-extruded films like LDPE/HDPE/LLDPE or HDPE/HM-HDPE, laminates such as BOPP/LDPE or Polyester/LDPE are being used.

11.7.2. Packaging of Extruded Products

Extruded products are classified into two categories traditional extruded products like sev, chakli, boondi etc. and non-traditional products made by industrial extruders such as kurkure, Cheetos etc. These extruded products are often enrobed by the oil, spice and by flavour compound. Crispiness of the extruded products is one of its characteristic should be preserved in the packaging. As it contains spice and flavour compounds the packaging material should be barrier to the volatile compounds to retain its flavour. Moreover, oil in the extruded will undergo oxidation, thus should be also taken in consideration for the packaging of the extruded products. LDPE or PP films are widely used for the packaging of local/traditional the extruded products. For branded extruded products laminates like BOPP / LDPE, BOPP / Polyester / LDPE, BOPP / Metallised Polyester / LDPE, Polyester / LDPE are being used.

11.7.3. Packaging of Bakery Products

Bakery products includes bread, cake, cookies and biscuits etc. These bakery products are classified as dry (biscuit and cookies etc.) and moist bakery products (bread, buns and cake etc.) based on their moisture content.

Dry bakery products are characterised by the moisture content of less than 6% and water activity of 0.30. These dry bakery products are hygroscopic which will intake the moisture easily. This hygroscopic nature of the dry bakery products significantly affects the organoleptic properties of the bakery products. Along with that, dry bakery products like cookies has high amount of fat so that they will readily undergo rancidity.

Initially, dry bakery products are packed in the paper cartons and polypropylene these days they are replaced by the flexible packaging with laminates. A mostly used flexible packaging material for dry bakery products is Cellophane. This cellophane film provides better barrier properties and heat saleability. This cellophane film is expensive and replaced by BOPP. For high quality dry bakery products combinations of OPP like OPP/PE, OPP/PET are being used for the packaging. Thermoformed trays made up of PVS and PS are also used with active substance like moisture and oxygen absorbers to increase the longevity of the dry bakery products.

Moist bakery products like cake and bread have moisture content of more than 12% and water activity range between 0.65 to 0.85. Staling is the one of the spoilage factor of the moist bakery products. Staling is related to starch retrogradation defined as recrystallization of gelatinised starch. This retrogradation of starch expels the water from the polymer network of bread. Because of its moisture content, moist bakery products are more prone to the microbial growth. Other spoilage mode of the moist bakery products includes oxidation, enzymatic reaction and loss of flavour. Conventionally these moist bakery products are packed in the waxed paper. Nowadays, PE films such as LDPE, LLDPE and PP films are used as cost effective alternative for the packaging of moist bakery products. Other packaging material used to pack the moist bakery products includes PVC, PP etc.

11.7.4. Packaging of Snack Foods

Most of the snack products are fried products includes chips and other savouries. Apart from the fried food nuts also comes under the classification of snack foods. Similar to the other fat containing foods, snack foods should also be protected from oxidative and hydrolytic rancidity and moisture migration. Absorption of moisture leads to the loss of the crispiness of the snack food products. It should be noted that packaging material for the snack foods should be barrier to oxygen and moisture, grease resistance and resistance to odour permeation. Moreover, these products are more prone to physical damage as they are so brittle and crisp in nature.

LDPE and PP pouches are used for traditional snacks with low shelf life. The laminates such as metallised polyester/LDPE, BOPP/LDPE Polyester/Al foil/ LDPE have been used for packaging of snack product for longer shelf life. Waxed glassine paper coated with cellophane also used for the packaging of the snack. Typically MAP packaging employed to extend the shelf life of fat rich snack food like chips and other fried foods. Gas composition of 100% N_2 is commonly used used in MAP. It should be mentioned that gas flushed MAP also protects the snack food from the physical damage as well.

11.7.5. Packaging of Confectionery

Confectioneries are food items that are rich in sugar and carbohydrates. Hard boiled sweets, medium hard sweets, chewing gum and chocolate come under the umbrella of the confectionery. They should be protected from the dust, dirt and infestation from the insects. Other spoilage modes are the colour and flavour loss. Some of the confectioneries become unpalatable due to the moisture migration. This moisture migration cause change in stickiness, hardness and desiccation of the confectioneries. Confectionery which contains fat are susceptible to the oxidation. Other defects of confectioneries includes fat bloom , sugar bloom and odour absorption. Water vapour transmission rate, gas transmission rate, tear strength, grease resistance are the preferred characteristic of packaging materials for confectioneries.

Hard boiled sweets contains moisture content of 3% and these are packed in waxed paper, PP, PVC, cellophane and HM-HDPE. Toffee and caramel are the example for medium hard sweets and their moisture content range from 8%-15% . These medium hard sweets are packed in the moisture proof cellophane. Wax paper with twist wrap are also being used. Chewing gum is being packed in laminates such as paper/PE or cellophane/PE. Chocolates are packed in inner side aluminium foil/ waxed paper and outside metallised PET/PE.

11.8. Packaging of Beverages

11.8.1. Packaging of the Drinking Water

Unlike other food products, water rarely undergoes spoilage as water is subjected to RO treatment and subsequent aseptic packing. Water don't have nutrient for the growth of microbes that mean carbon and nitrogen source are one of the critical factors for microbial growth. However, water should be protected from physical, chemical and microbial contamination during storage and transportation. RO (Reverse Osmosis) treated water are packed either in rigid bottles or in flexible polymer packaging. Some of the requirements for the packaging of the drinking water according to FSSAI includes, the packaging material should be transparent and tamper proof. Transparent Polypropylene, Polyvinyl chloride, polyethylene terephthalate are used for the rigid packaging of water. Low density polyethylene films are used for the flexible packaging of the water. Another important consideration is that these polymers should not migrate and alter the odour of water which reduces the shelf life of the packaged drinking water.

11.8.2. Packaging of Soft Drinks

Soft drinks are often referred to the carbonated drinks that are usually made up of water, sweeteners and flavourings. Rigid containers are mainly used in the packaging of soft drinks as it is able to withstand the high partial pressure that is exerted by carbon-di-oxide inside the package without distortion and expansion. Along with the packaging material should have low permeability for the CO_2 gas and has to maintain the carbonation of the beverage. Another mode of spoilage is hydrolytic and oxidative rancidity of flavouring oils. In the aspect packaging, oxidative rancidity is our interest as it facilitated by the oxygen. CO_2 is more permeable than O_2 as it has more solubility means selection of the packaging material with low CO_2 permeability will automatically fulfil the low O_2 permeability. This is only applicable for oil based flavouring substances. Migration of monomers from the packaging will form undesirable flavour or odour in the soft drink which reduces the acceptability of the beverage. Antioxidant additives like arlyamines, phenolics are used to reduce the migration of monomer from the plastic resins.

Widely used packaging materials for the soft drinks are blow moulded and bi-axially stretched PET bottles, three piece tinplate or two piece aluminium metal cans and glass bottles. Glass bottles with crown closure are the better choice as it has properties like low permeability and ability to withstand high internal pressure for the packaging of the carbonated beverage. In the packaging of these carbonated drinks headspace should be given to avoid the expansion or distortion caused by the development of high internal pressure due loss of CO_2 solubility in water. For secondary packaging either corrugated fibre- board boxes or polymer shrink wrap is used.

11.8.3. Packaging of Coffee

Coffee available in the market can be classified as roasted whole coffee beans or roasted ground coffee beans and instant coffee which is readily soluble prepared by spray drying. Ground coffee and instant coffee are more likely to lose its aroma than whole coffee beans which will reduce the quality of ground coffee beans. Moreover, Instant coffee and Ground coffee tends to absorb moisture which clumps which leads to the spoilage. Fatty acid in coffee beans undergoes the rancidity which significantly affects the acceptability. The major spoilage of coffee is due to the staling of the coffee. The staling of the coffee refers to the oxidation of linoleic triglyceride and causes the unpleasant flavour of the coffee. So that it is necessary to keep the coffee out of oxygen to maintain its fresh aroma. Another vital consideration while choosing the packaging material for the roasted coffee is the CO_2 permeability. Because, roasted coffee beans release the CO_2 that should be vented out from the package to avoid the expansion of the package. In conclusion, the coffee should be kept out light, moisture and mainly from oxygen to increase the shelf life of the coffee.

Conventionally, vacuummized tinplate metal cans were used for packaging of the roasted and ground coffee. Nowadays, manufacturers are using multilayer laminates like PET/Al foil /LDPE for the packaging of the Roasted ground coffee. Sometime metallized PET/LDPE laminates also used as alternative for PET/Al/LDPE. These Multi-film laminates are sealed hermetically to ensure the air tight packaging. Roasted whole or ground coffee are usually packed in vacuum or modified atmospheric condition (100% N_2). Oxygen absorbers are also placed inside the package to eliminate the residual oxygen inside the package. While considering the single polymeric packaging, PET shows the better result than LDPE, PVC, PS.

Instant coffee is packed in laminates like PET/Al foil/LDPE or metalized PET/LDPE, metal cans and glass containers. These packaging material act as very good barrier to moisture which is the major deteriorative factor in the instant coffee so that instant coffee can be stored for longer period of time than roasted whole and ground coffee. The reason behind is that some of fatty and flavour substance that cause for the staling of the whole or ground coffee beans is absent in the instant coffee

11.8.4. Packaging of Tea

Different types of tea are present in the market includes white, green, oolong, black tea. These types of tea are based on the level of the fermentation tea leave and other processing factors during manufacturing process. Regardless of type of tea, all tea variants contain various volatile aroma compounds and essential oils providing unique flavour and aroma for the different types of tea. Polyphenol like Catechins are responsible for the flavour which is predominant in white and green tea. These polyphenols are more prone to the oxidation that is why green and white tea has less shelf life than black tea and other types of tea. Observed changes during spoilage of green tea are reduction in ascorbic acid due to oxidation and colour change of tea from green to olive. Vacuum packaging and gas flushing are employed to evacuate the oxygen responsible for oxidation in the green tea packaging. In contrast, black and oolong tea are rich in compounds like theaflavins and thearubigins responsible for the colour and aroma. These compounds should be protected from aroma loss to preserve its quality and freshness. Along with it is vital to maintain the optimum moisture content of tea. This can be achieved by using packaging material with less water vapour permeability.

Metals have less of permeability to water vapour, oxygen as well as volatile compounds. Some premium tea manufacturers are using the metal containers for the packaging of the tea. Most commonly used packaging for retail includes Paperboard/Al foil or Paperboard /PP laminate. Other laminates used for the packaging of the tea includes LDPE, paper coated LDPE, PET/LDPE and BOPP/LDPE. Nowadays, tea bags are becoming popular among consumer those kinds of tea bags are packed in tissue paper with high amount of pore and higher wet- strength. Researches are still in development in the innovative packaging of tea bag.

11.8.5. Packaging of Fruit Juices

Fruit juices are non-carbonated beverages made from fresh fruits contains invert sugar, fruit extract, dextrose and water etc. Clarified juice, nectar, marmalade and squash comes under the classification of the fruit juices. Fruit juices have acidic pH so that they are susceptible to the spoilage by yeast rather than bacteria. To prevent this juice is pasteurized before packaging. However, other deteriorative factors of fruit juice during storage includes,

- Reduction in ascorbic acid content: due to the presence of oxygen, ascorbic acid gets self-oxidized by ascorbic acid oxidase.

- Loss or degradation of volatile flavour substances caused by the migration of monomers from the packaging material.
- Non-enzymatic browning facilitated by reaction between the amino acid and reducing sugar mostly observed in citrus juices.
- Loss of colour due the oxidation of colour responsible compounds like carotenoids and anthocyanin etc.

These are factors should be considered in the packaging of the fruit juices. The packaging material should be less permeable to water vapour and oxygen. In addition to that most of the fruit juices are hot filled after pasteurization of juice so that the packaging material must be temperature resistance.

Traditionally, Lacquered tin metal cans were used for the packaging of the fruits juices. As fruit juices are acidic in nature, it will cause the corrosion of the metal cans through the scratches in the metal cans. Glass containers are also used for its inert nature and its high barrier property to oxygen & volatile substance present in the juice. However, the care should be taken while hot filling the glass container as glass containers are more prone to thermal shock. Still, glass containers are in use for the high quality fruit juices packaging. Nowadays fruit juices are packed in the laminates like tetra pak (PE/PE/Al foil/PE/Paper board/PE). These laminates are disinfected using UV treatment that facilitates the asceptic packaging of the fruit juices. Rigid PET containers are also used widely in the market due to its thermal properties and barrier properties for the packaging of the fruit juices. PET has more glass transition temperature that means it is more resistance to temperature so that it can be used for the hot filling of fruit juices .

11.8.6. Packaging of Beer

Beer comes under carbonated alcoholic beverages prepared from brewing of malted barley. Hop is bitter flavouring substance used in the beer. It has pH of 4.0 slightly acidic. Due to its high acid content and aseptic packaging condition(pasteurization of beer and disinfection/ sterilization of packaging of material),beer is less susceptible to microbial contamination. CO_2that is formed during the fermentation of the beer should be maintained just like carbonated soft drinks to maintain its quality. It means the factors that are considered in the packaging of soft drinks are somewhat similar in packaging of the beer as well. However, other potential spoilage modes of beer include haze formation and flavour loss of beer.Haze formation is not controllable factor through the packaging, as it is facilitated by the organic compound like proteins, polyphenols that is present in the beer itself.

Usually bottles with crown closure and metal cans with pull tabs at the top of the metal cans are used for the packaging of the beer. Glass bottle is inert so that it has no effect on the flavour the beer. Considering the cost of glass bottles, tinplate metal cans are also used for the packaging. These metal cans are lacquered to reduce the chemical interaction of beer with metal containers and leaching of metallic flavours from the cans. Low CO_2 permeability and temperature resistance of glass bottles and metal cans makes them suitable for the beer packaging. Sometimes, rigid plastic containers made up of PVC/PVdC copolymer-coated PET bottle is used for the packaging of the beer. PVC/PVdC copolymer coating in PET bottles are used to improve the oxygen barrier and for the prevention of the beer flavour. Wooden boxes and corrugated fibre board boxes are used for secondary packaging of the beer for the transit purpose.

11.8.7. Packaging of Wine

Wine is an alcoholic beverage made from grapes by the yeast fermentation. Major spoilage of wine is the oxidation which leads to the browning reaction and development of unpleasant flavours. This oxidation of wine changes the characteristic taste of wine.

Coloured glass bottles with cork or screw caps are widely used for packaging of wine. This inert glass containers doesn't alter the flavour and blocks the light to prevent the oxidation. Another alternative packaging for the packaging of wine is that bag in box type packaging. Usually these bags are made by the co-extrusion of EVA-BA-EVOH-BAEVA or LDPE-BA-EVOH-BA-LDPE. PET bottles are also used as alternative for glass containers for shorter shelf life of wine. In PET bottles upon longer storage time, oxygen may enter inside the package which significantly changes the flavour of wine and changes the solubility of some components present in the wine which cause clouding and sediment of wine. So , PET bottles are not intended to use for longer storage of wine. Along with that cartons like tetrapak are also used. However, Glass bottles are considered as recyclable and good packaging material for the packaging of the wine.

11.9. Packaging of Flesh Foods, Poultry and Seafood

The term 'flesh foods or meat' refers to muscle of warm blood four leg animals like cattle, pig and sheep.Poultry refers to meat of the domesticated birds like ducks, chicken, turkey, ostrich etc. Whereas fish and shellfish like squid, octopus, crustaceans comes under seafood. These foods from animal origin are consumed as fresh or cured or processed into different products like sausage. This section will cover the spoilage modes of those products and the appropriate packaging material and packaging technique to reduce the spoilage of different products.

11.9.1. Packaging of the Fresh Meat and Poultry

Fresh meat and poultry are more perishable than cured meats. As they have high moisture content, they are susceptible to microbial growth. Another mode of the spoilage of fresh meat and poultry is moisture loss which leads to weight loss, shrinkage and reduce the tenderness, juiciness of the meat. Colour of the meat also impacts the acceptability of fresh meat. Availability of O_2 also promotes the growth of the aerobic microorganisms like Pseudomonas species. In contrast, Formation of oxymyoglobin (bright red colour) from myoglobin requires oxygen. Thus oxygen content in the meat packaging should be maintained in such way to preserve its colour as well as control the microbial growth. Also fresh meat have high amount of fat which tends to absorb odour from the surrounding, thus affects the odour of the meat. All these factors should be considered in the selection of proper packaging material in order to maintain the quality of fresh meat and poultry.

Foamed polystyrene or transparent PVC trays with overwrap are used to pack fresh meat. These trays are covered by the LDPE or PVC. Shrink film overwrap also used to pack the fresh meat product. In this method, fresh meats are placed in conveyer that goes into the tunnel where hot air is blown over the polymers so that wrapping of fresh is achieved. For the shrink wrapping of the fresh meat products PVC, cellophane, rubber hydrochloride, polypropylene, irradiated PE and PVDC are used.

Vacuum packaging using Polyester/PE film laminates, Polyamide/PE film laminate, Aluminium foil/PE laminate, PVDC/PE laminate and PVDC copolymer in combination with refrigeration is used for the longer storage of the packaging of fresh meats and poultry. This eliminates the oxygen and minimizes the deterioration by inhibiting the growth of aerobic organisms and preserving the red colour of the meat. MAP with gas composition of 20-30% CO_2 and 70-80% O_2 is used for packaging of fresh meat. CAP is used for preservation of the whole carcass like beef (not for retail and individual packaging) which uses nitrogen to provide the anaerobic condition, CO_2 to provide the inhibitory effect over microbes and control over amount of oxygen depending on the meat product .For example, Colour of beef carcass is important characteristic so that it uses high amount of O_2, whereas pork carcass contains high amount of fat so that it uses less O_2 in its atmosphere. However, in the most cases, anaerobic atmosphere is used in the CAP indefinitely for the packaging of red raw meat which is similar to vacuum packaging in terms of longer shelf life of meat and its lower O_2content in their packaging atmosphere.

It should be noted that the colour of raw red meat is not good in the vacuum and anaerobic CAP due to the lack of oxygen in their packaging atmosphere.

In recent advancements, antimicrobial packaging which contains antimicrobial agents like chitosan, tea tree oil, cinnamon, thyme etc. and intelligent packaging like TTI (time temperature indicator) which indicates the freshness of the products are used in the packaging of the fresh meat and poultry.

11.9.2. Packaging of the Cured Meat and Cooked Meat

Curing of meat is used for the preservation with curing agents like nitrates, nitrites and sodium chloride. Cured meat has shelf life of 10 weeks, if it is stored at 0^0C. During this curing process, pigments in the meat undergo certain chemical reactions and forms the nitrosomyoglobin which is pink in colour. These salts reduce the water content of the fresh meat by the phenomena called osmotic dehydration. Therefore, curing agents inhibits the growth of the microorganisms as well as improve the colour of the meat. Flavour development also considered as a role of curing. This flavour of the cured meat should be preserved during the storage. These cured meats are more susceptible to the oxidative rancidity. Formerly mentioned characteristics of the cured meat are significant factors for the selection of the packaging material and packaging techniques. For the short term storage of cured meat, LDPE is used for its high barrier properties. Whereas for long term storage of cured meat,laminated vacuum packaging is employed. MAP with gas composition of 85-90% $N_2$and 10-15% CO_2 is used for the longer shelf life of the cured meats.2:1 is the preferred gas to product ratio for the MAP of the cured meat. This gas mixture ensures the prevention of the oxidative rancidity of cured meat.

Cooked and processed meats such as sausages, nuggets and meat balls. For short term storage these are stored in the pouches of polyethylene, polypropylene, PVDCetc. Sometimes, meats are canned aseptically in sterile conditions and then cooked in the can itself. This can be stored for 2 years at ambient temperature. However, retort packages are also used to pack the cooked meat.

11.9.3. Packaging of the Sea Food

One of the major deteriorative reactions of fish and shellfish is autolysis of fish during the storage of sea food at lower temperature. Autolysis involves the breakdown of fish protein and this process is facilitated by the enzyme that is present in the fish and shellfish. Oxidation and microbial growth or contaminations are the other possible modes of the fish and shellfish spoilage. Autolysis cannot be controlled by the means of the packaging as it caused by the fish enzyme itself. In the other hand, aerobic microbial growth and oxidative spoilage can be controlled by reducing the amount of the oxygen inside the packaging. Oxidative reaction possess more threat for sea foods than flesh meat, as sea food are rich in PUFA (Poly unsaturated fatty acid). In addition to that packaging materials should be less to permeable to odour and water vapour. So that odour permeation and drip loss can be prevented. Insulation properties of the packaging material another important factor while choosing packaging of sea foods as they are stored and transported in the frozen condition.

Conventionally, wooden crates are used for the bulk transportation of fishes. Now, Foamed polystyrene or moulded polystyrene trays also used for the packaging of the sea food for their better insulation properties and light weight to facilitate the bulk transportation.Aluminium containers also used as they are providing better insulation but they are heavy in terms of the weight.In the markets shallow, clear polystyrene trays are used for retail purpose. These trays are mostly double walled for the better insulation and overwrapped with PVC films. However, this is not common in the India.MAP will have significant effect on the shelf life improvement of sea food. For sea food with less or non-fat, 30% O_2, 40% CO_2 and 30% N_2 gas composition is used in MAP. CO_2in MAP inhibits the growth of microbes because of it bacteriostatic effect. Fat rich sea food and smoked sea food uses MAP gas composition of 40%, CO_2 with 60% N_2.

The following are the important point for selection of packaging material for MAP system

- High gas barrier films like PVdC co-polymers to maintain the gas composition inside the package.
- High water vapour barrier films like PVdC and HDPE to prevent drip loss.
- Heat sealable films like LDPE, EVA that facilitates the tight seal between lid and tray to prevent moisture loss and microbial contamination.
- Oil resistance films like PVDC and polyamides can be considered for fat rich sea foods.
- Transparency of films is necessary to observe the quality of the fresh sea food.

Drip absorbers are placed in the bottom of the MAP to absorb the drip caused by the fresh packaged sea foods. This will reduce the bacterial growth by provide clean package environment. Example for the drip absorber is Cellulose pads.

IQF (Individually Quick Frozen) shrimps are packaged in LDPE Pouches, also Polylaminated cartons like Co-extruded LLDPE/B-Polyamide/B-LLDPE (five layered film) , Polyester / LDPE are used for packaging of sea food. Individual fish fillets are also packaged in cellophane or PVC bags.

References

Barbosa-Cánovas, G. V., Altunakar, B., &Mejía-Lorío, D. J. (2005). Freezing of fruits and vegetables: An agribusiness alternative for rural and semi-rural areas (Vol. 158). Food & Agriculture Org..

Bhattacharya, G. (2003). Trends in Food Packaging for Organised Retail. Packaging India, 35(5), 69-72

Blakistone, B. A., & Blakistone, B. A. (Eds.). (1998). Principles and applications of modified atmosphere packaging of foods. London: Blackie Academic & Professional

Brody, A. L., Bugusu, B., Han, J. H., Sand, C. K., & McHugh, T. H. (2008). Innovative food packaging solutions. Journal of food science, 73(8), 107-116.

Catsberg, C. M., &Kempen-van Dommelen, G. J. (2013). Food handbook. Springer Science & Business Media.

Cenci-Goga, B. T., Iulietto, M. F., Sechi, P., Borgogni, E., Karama, M., &Grispoldi, L. (2020). New Trends in Meat Packaging. Microbiology Research, 11(2), 56-67.

Coles, R., McDowell, D., &Kirwan, M. J. (Eds.). (2003). Food packaging technology (Vol. 5). CRC Press.

Dordi, M. C. (1998). Modern food packaging.

Farzana, E. W. Packaging of Snack Foods.

Franzetti, L., Martinoli, S., Piergiovanni, L., &Galli, A. (2001). Influence of active packaging on the shelf-life of minimally processed fish products in a modified atmosphere. Packaging Technology and Science: An International Journal, 14(6), 267-274.

Ghose, P., & Nair, P. (2013). Packaging of carbonated beverages. Int. J. Agric. Food Sci. Technol, 4(5), 421-430.

Gopirajah, R., &Anandharamakrishnan, C. (2017). Packaging of dried dairy products. Handbook of Drying for Dairy Products, 229-248.

Jenkins, W. A., & Harrington, J. P. (1991). Packaging foods with plastics.

Jeyasekaran, G., Jeya Shakila, R., &Sundararaj, V. (2002). Opportunities in fish processing. Indian food industry, 21(6), 40-42.

Kiyoi, L. (2010). Determining the Optimal Material for Coffee Packaging: Oxygen ransmission Rates and Ink Abrasion Resistance.

Klimchuk, M. R., &Krasovec, S. A. (2013). Packaging design: Successful product branding from concept to shelf. John Wiley & Sons.

Kontominas, M. G., Badeka, A. V., Kosma, I. S., &Nathanailides, C. I. (2021). Recent Developments in Seafood Packaging Technologies. Foods, 10(5), 940.

Kumar, K. R. (2002). Packaging aspects of fruit beverages. Beverage and Food World, 29(9), 30-32.

Mangaraj, S., &Goswami, T. K. (2009). Modified atmosphere packaging of fruits and vegetables for extending shelf-life-A review. Fresh produce, 3(1), 1-31.

Matche, R. S. (2005). Packaging aspects of fruits and vegetables. Plastics in food packaging. Indian centre for plastics in the environment, 115-132.

Miranda, G., Berna, A., &Mulet, A. (2019). Dried-Fruit storage: An analysis of package headspace atmosphere changes. Foods, 8(2), 56.

Miranda, G., Berna, A., &Mulet, A. (2019). Dried-Fruit storage: An analysis of package headspace atmosphere changes. Foods, 8(2), 56.

MirzaAlizadeh, A., Masoomian, M., Shakooie, M., ZabihzadehKhajavi, M., &Farhoodi, M. (2020). Trends and applications of intelligent packaging in dairy products: a review. Critical Reviews in Food Science and Nutrition, 1-15.

Morata, A. (Ed.). (2018). Red wine technology. Academic Press.

Motey, R., &Lele, S. (2003). Plastic films for processed foods-special requirements. Packaging India, 35(5), 19-32.

Paine, F. A., & Paine, H. Y. (2012). A handbook of food packaging. Springer Science & Business Media.

Paramasivan, Karthickumar& Moses, J.A.. (2015). Vacuum packaging of meat. Food Marketing and Technology. 42-43.

Potter, N. N., & Hotchkiss, J. H. (1995). Principles of food packaging. In Food Science (pp. 478-513). Springer, Boston, MA.

Pouch, L. P. Packaging of Seafood.

Prabhu, P. V., & Gopal, T. S. (1998, January). Modern trends in seafood packaging. Cochin University of Science and Technology.

Robertson, G. L. (2005). Food packaging: Principles and Practice. CRC press.

Robertson, G. L. (2016). Packaging of beverages. In Food Packaging (pp. 607-636). CRC Press.

Sacharow, S., & Griffin, R. C. (1980). Principles of food packaging. AVI Pub. Co..

Ščetar, M., Kurek, M., & Galić, K. (2010). Trends in fruit and vegetable packaging–a review. Hrvatskičasopiszaprehrambenutehnologiju, biotehnologiju i nutricionizam, 5(3-4), 69-86.

Shankar, C. N. R. (2005). Packaging of value added fish products. Central Institute of Fisheries Technology.

Snel, M. A. R., Bakker, R. W. M., &Berkenbosch, A. C. (1998). Packaging of fish. Modern Food Packaging, 412-430.

Soltani, M., Alimardani, R., Mobli, H., & Mohtasebi, S.S. (2015). Modified atmosphere packaging: a progressive technology for shelf-life extension of fruits and vegetables. Journal of Applied Packaging Research, 7(3), 2.

12

Food Packaging Laws

Subhamoy Dhua[1], Pravin M. Ganorkar[2] and Kshitiz Kumar[2]

[1]*Department of Food Engineering and Technology, Tezpur University Assam, India*
[2]*A.D. Patel Institute of Technology, CVM University, V V Nagar, Gujarat, India*

12.1. Introduction

Food Packaging is a means to ensure safety and quality of food product throughout storage, transportation, handling, distribution, retailing, sales and in some cases till the intended use. International trades in food are generally guided by Codex Alimetarius Commission (CAC) (A joint body of World Health Organization and Food & Agricultural Organization). CAC works in harmonization within the context of food standards, food packaging and the applied rules. The mandate of Sanitary and Phytosanitary Measures (SPS Agreement) and Agreement on Technical Barriers to Trade (TBT Agreement) is to facilitate food trades and remove restraints at possible extent. Through these agreements, member countries have provision to apply quality and safety measures to safeguard the consumer. Such measures extend to every aspect of food and its packaging. On the other side, food packaging industries also progressed with sound technological advancement and sophistication in the last three decades. Innovations in food packaging attracted consumers due to convenience, availability, safety, quality and preservation throughout shelf life of the food product. The basic function of food packaging is to protect the food from deterioration and contamination. Moreover, the important function of food packaging is to carry labelling information which will make aware about the product to the consumer. Therefore, food packaging becomes integral part of the food product. If food packaging fails in the stated functions, significant negative impact on the food trade is likely to occur. This may have impact horizons on consumer's safety. As per Code of Federal Regulation (21 CFR parts 174-178), packaging materials are considered an 'indirect food additive'. They are defined as materials that come into contact with food as part of packaging, holding, storage or processing. The packaging material that contacts food shall be of purity suitable for its intended use. Food that is not packaged according to regulations will be considered adulterated and prohibited from being sold. Therefore, it is a vital requirement to synchronize the efforts of CAC and national level bodies towards consumer's safety and at the same time facilitating trades without any barriers or conflicts. In context of this, important features of SPS Agreement, TBT Agreement and Indian Food Packaging laws are discussed in this chapter.

12.2. SPS Agreement and TBT Agreement

Codex Alimentarius Commission, the Joint FAO/WHO Expert Committee on Food Additives (JECFA) and through technical cooperation with country members, play a role in promoting and assuring adequate levels of food quality and safety, in protecting consumers, improving domestic food supplies and in reducing costly rejection of foods shipped in international trade. The purpose of agreements is to have optimum tariffs, reduce the impact of non-tariff trade barriers and eliminate tariffication measures like discretionary licensing, variable levies, quotas, restrictive border measures minimum import prices, voluntary restraint agreements. Equivalency, harmonization and transparency are vital in the World Trade Organization (WTO) Agreement on the Application of Sanitary and Phytosanitary Measures (SPS Agreement) and the Agreement on Technical Barriers to Trade (TBT Agreement).

The SPS Agreement primarily affirm that importing countries can enforce process obligatory to guard the life or health of their citizens, along with the life and health of their animals and plants, on the basis of sound scientific evidence and judgment, using suitable risk analysis methods. The SPS Agreement provides procedures for dispute settlement. The SPS Agreement established customary WTO committee on Sanitary and Phytosanitary Measures to provide a regular forum for consultations and to monitor the process of international harmonization and the use of international standards, guidelines and recommendations.

The TBT Agreement is a revision of the agreement first developed under the Tokyo Round of GATT Negotiations in the 1970s. It focuses on the elimination of the use of national technical requirements, or standards in general, as unjustified technical barriers to trade. The TBT Agreement principally provides that all technical standards and regulations must have a valid intention and that the impact or cost of implementing the standard must be proportional to the purpose of the standard. It also says that if there are two or more ways of achieving the same purpose, the least trade-restrictive substitute should be followed.

The Codex Alimentarius Commission adopted a Code of Practice in 1995 for packaging and transport of tropical fresh fruit and vegetables. The provisions in this code, consistent with the approach of the Uruguay Round Agreement and the principles of equivalence, set out standards to be achieved rather than prescribing detailed requirements. Packaging must withstand rough handling during loading and unloading, compression from the overhead weight of other containers, impact and vibration during transportation, and high humidity during precooling, transit and storage. The packaging components or monomer or traces should not migrate to the food. If it gets migrated, it should be within maximum permissible limits specified by regulatory bodies. For example, CAC depicted guideline levels for vinyl chloride monomer (0.01mg/kg) and acrylonitrile (0.02mg/kg) in food with slightly higher levels permitted in packaging material. Moreover, the application of 'Good Manufacturing Practices' must be employed with the advancement of food packaging. The hazards associated with any particular new combination of food and packaging are scrutinized in advance, and measures taken to control the hazard so as to ensure consumer protection.

Moreover, ISO standard TC 122 deals with packaging in general. The ISO/TC 6 which deals with paper, board and pulp has 186 standards formulated under this category. For light gauge metal containers there is ISO/TC 52 and ISO/TC 63 deals with glass containers under which ISO has formulated 32 standards.

12.3. National Level Food Packaging Laws

The Indian Government has enacted laws to take care of quality standards of packaged foods. Standards have also been fixed for the particular kind of packaging required to be undertaken, depending on the product to be packed. The regulations on quality standards of packaging that govern food products in domestic market are given as under.

12.3.1. Standard Weights and Measures Act (SWMA) 1976 and the Standards of Weight and Measures (Packaged Commodities) Rule, 1977

Ministry of Consumer Affairs, Food and Public Distribution governs 'Standards and Measures Act' (SWMA) in India. It is mandatory and applicable to all packed commodities including food products. Indian Institute of Legal Metrology provides training in legal metrology to inspectors and other interested parties. SWMA focus on packed goods quantity and price declaration on the label to assist comparisons and protect consumer welfare. It prescribe specification of measuring instruments used in commercial transaction, industrial production and measurement involved in public health and human safety. It also provides control and regulation of export and import of weights and measures and commodities in packaged form. The standard specifies quantities to be packed, expressions to be avoided and size of type depending on the size of the panel in a package. The SWMA requires certain declaration to be made on every retail package, which includes common/ generic name of the product, net quantity, retail sale price, unit sale price, month and year of manufacturing or pre-packing, and name and address of the manufacturer or the packer. As far as possible, all declarations required to be made under SWMA should appear on the principal display panel (PDP) of the package. SWMA empowers inspectors towards inspection of weighing and measuring instruments during their use in commodities packaging industries to prevent fraudulent practices.

12.3.2. Agriculture Produce Grading and Marking (AGMARK) Act , 1937

AGMARK Act execution is carried out by The Directorate of Marketing and Inspection (DMI), Department of Agriculture, Cooperation and Farmers Welfare under Ministry of Agriculture & Farmers Welfare. The Directorate is headed by Agriculture Marketing Adviser to Government of India and has its head office at Faridabad (Haryana), branch head office at Nagpur (Maharashtra). AGMARK is quality certification mark for agricultural products and act as third party guarantee to certified quality. Agricultural products such as cereals, pulses, nuts, fruits & vegetables, bread wheat flour, ghee, essential oil, honey, spices and condiments, edible oils, livestock, fiber crops etc. are covered under AGMARK for their quality parameters. AGMARK facilitates products grading. The product categories like regular, standard and premium are generally offered. Manufacturer can set higher prices for premium quality products. On the other hand consumers are assured that they are purchasing premium quality product when paying higher prices. The AGMARK rules also specify the type of packages that can be used and labeling declarations that has to be given. It is voluntary for internal trade and compulsory for export oriented products. AGMARK logo is permitted to print on the package or label after product certifications.

Fig. 1: Logo of AGMARK

12.3.3. Bureau of Indian Standards (BIS) Act, 1986 and BIS Rules, 1987

BIS is a statutory institution established under the Bureau of Indian Standards Act, 1986 to promote harmonious development of the activities of standardization, marking and quality certification of goods and attending to connected matters in India. Each activity is carried out by the respective department. Sectional committees are responsible to a specific product category. BIS certification is mandatory for 16 food products like follow–up formula –complimentary foods, processed cereal based complementary foods, milk-powder, partly skimmed milk powder, condensed milk, sweetened ultra high temperature treated condensed milk, skimmed milk-powder, standard grade and extra grade, packaged drinking water, packaged natural mineral water, infant milk substitutes, milk – cereal based complementary foods, plastic and glass feeding bottles. The BIS has formulated specifications for packaging materials, packages and components. Four important packaging materials i.e. paper & cardboard, plastic films in single or multi-layer laminates or co-extruded films, metal containers and glass containers, are used widely in food packaging. BIS specifies the types of packaging materials that can be used for various types of food products. These specifications are voluntary for most of the foods, but are compulsory for certain items like food colours and packaged drinking water. There's a possibility of migration of part of packaging material into the food items due to contact. It is vital that package should be selected on the basis of minimum migration. BIS standard IS 9845 is for determining overall migration of constituents of plastic and articles intended to come in contact with foodstuff. Another standard IS 9833 lists the permitted pigments and colorants that may be used in plastics that may come in contact with foodstuffs, pharmaceuticals and drinking water. BIS IS 10171 depicts the suitability of plastics for food packaging to help manufacturers. As per Plastic Waste Management rules, 2016, carrybags or products made of recycled plastic shall not be used for carrying, storing, dispensing or packaging of food items.

12.3.4. Food Safety and Standard Authority of India (FSSAI) – Packaging Regulations

Food Safety and Standard Authority of India Act, 2006 received the assent of the President on 23rd August, 2006. This act is to consolidate the laws relating to food and to establish the Food Safety and Standards Authority of India for laying down science based standards for articles of food and to regulate their manufacture, storage, distribution, sale and import to ensure availability of safe and wholesome food for human consumption and for matters connected therewith or incidental thereto. Presently there are 22 Food Safety and Standards Regulations. In 2011 FSSAI made the Food Safety and Standards

(Packaging and Labelling) Regulations, 2011. In 2018 FSSAI decided to divide these regulations into two regulations,-

(i) Food Safety and Standards (Packaging) Regulations, 2018; and

(ii) Food Safety and Standards (Labelling and Display) Regulations, 2018,

This chapter covers the regulations covered under the Food Safety and Standards (Packaging) Regulations, 2018 while regulations related to labelling and display are covered in Chapter 13.

12.4. General Packaging Requirements

(i) Every food business operator shall ensure that the packaging material used shall be in accordance with the Food Safety and Standards (Packaging) Regulations, 2018

(ii) Any material which comes in direct contact with food or likely to come in contact with food used for packaging, preparation, storing, wrapping, transportation and sale or service of food shall be of food grade quality.

(iii) Packaging materials shall be suitable for the type of product, the conditions provided for storage and the equipment for filling, sealing and packaging of food as well as transportation conditions.

(iv) Packaging materials shall be able to withstand mechanical, chemical or thermal stresses encountered during normal transportation. In case of flexible or semi-rigid containers, an overwrap packaging may be necessary

(v) Food products shall be packed in clean, hygienic and tamper-proof package or container

(vi) The sealing material shall be compatible with the product and the containers as well as the closure systems used for the containers.

(vii) Tin containers once used, shall not be re-used for packaging of food.

(viii) Plastic containers of capacity 5 litre and above and Glass bottles, which are reused for packaging of food, shall be suitably durable, easy to clean or disinfect.

(ix) Printing inks for use on food packages shall conform to IS: 15495.

(x) Printed surface of packaging material shall not come into direct contact with food products.

(xi) Newspaper or any such material shall not be used for storing and wrapping of food.

(xiii) In case of multilayer packaging the layer which comes in direct contact with food or layers likely to come in contact with food shall meet the requirements of packaging materials specified in Schedule I, II and III of these regulations.

(xiii) The materials listed in Schedule I, II and III of these regulations shall be compatible with their intended use as a packaging material so as not to alter the quality and safety of the food product.

(xiv) Every food business operator shall obtain the certificate of conformity issued by NABL accredited laboratory against these regulations for the packaging material which comes in direct contact with food or layers likely to come in contact with food to be used.

12.5. Specific Packaging Requirements

12.5.1. Paper and Board Materials Intended to Come in Contact with Food Products

(i) Paper and board material shall be of uniform formation, thickness and substance.

(ii) It shall be free from visible specks, grease marks, cuts, pinholes and other blemishes.

(iii) The paper used for the manufacture of boxes, cartons, plates, cups and paper lids or paper which are meant to be direct in contact with food shall be of food grade and shall be free from contaminants.

(vi) Paper and board materials used for the manufacturing of containers for packing or storing the food products shall conform to either of the Indian Standards specifications as provided in Schedule – I.

Schedule – I Paper and board materials intended to come in contact with food products

S.No	Types of Paper	Indian Standards specifications to be conformed
1	Grease proof paper	IS 6622
2	Vegetable parchment or Grease proof paper or Aluminium Foil Laminate	IS 7161
3	Aluminium Foil Laminates for Packaging	IS 8970
4	General purpose packing or wrapping Paper	IS 6615
5	Folding Box Board, uncoated	IS 1776
6	Corrugated Fibre Board Boxes- Specification (Part 1)	IS 2771

Note: The wax used for coating the paper or board shall be paraffin wax conforming to Type I of IS 4654.

12.5.2 .Glass Containers Intended to Come in Contact with Food Products-

(i) As far as possible, they shall be free from blisters, mould marks, stones and chippings and as far as possible shall be free from cords, seeds and other visible defects.

(ii) They shall have a smooth surface without cracks, pinholes and sharp edges.

(iii) The sealing surface shall be free from hairline cracks and prominent seam marks.

12.5.3. Metal and Metal Alloys Intended to Come in Contact with Food Products

(i) A utensil or container made of the following materials or metals, when used in the preparation, packaging and storing of food shall be deemed to render it unfit for human consumption: -

- Containers which are rusty;
- Enamelled containers which have become chipped and rusty;
- Copper or brass containers which are not properly tinned.

(ii) Appropriate grades of metal and metal alloys where applicable shall be used for packing or storing of food products.

(iii) Metal and metal alloys used for the manufacturing of containers for packing or storing the food products shall conform to either of the Indian Standards specifications as provided in Schedule – II.

Schedule – II Metal and Metal Alloys intended to come in contact with food products

S.No	Types of metal and metal alloy	Indian Standards/ISO specifications to be conformed
1	Cold-reduced Electrolytic Tinplate	IS 1993/ISO 11949
2	Cold reduced Electrolytic Chromium or Chromium Oxide – Coated Steel	IS 12591/ISO 11950
3	Wrought Aluminium and Aluminium Alloy Sheet and Strip for General Engineering	IS 737/ISO
4	Aluminium and Aluminium Alloy Bare Foil for Food Packaging	IS 15392
5	Specification for Crown Closures	IS 1994
6	Specification for Round Open Top Sanitary Cans for Foods and Drinks	IS 9396 (Part 1)
7	Specification for Round Open Top Sanitary cans for Foods and Drinks	IS 9396 (Part 2)

12.5.4. Plastic Materials Intended to Come in Contact with Food Products-

(a) Plastic materials used for the manufacturing of containers for packing or storing the food products shall conform to either of the Indian Standards specifications as provided in Schedule – III

Drinking Water (both Packaged and Mineral Water) shall be packed in colourless, transparent and tamper-proof bottles or containers made of

(i) polyethylene (PE) conforming to IS: 10146 or

(ii) polyvinyl chloride (PVC) conforming to IS: 10151 or

(iii) polyalkylene terephthalate (PET and PBT) conforming to IS: 12252 or

(iv) polypropylene (PP) conforming to IS: 10910 or

(v) food grade polycarbonate conforming to IS: 14971 or

(vi) polystyrene conforming to IS: 10142 or

(vii) sterile glass bottles only.

The transparency of a container shall not be less than 85 percent in light transmittance:

Provided also that all pigments or colorants as specified in Indian Standard IS: 9833 may be allowed in plastic containers of five litre and above made of Poly Corbonate (PC) and Polyethylene Terephthalate (PET) used for packaging of mineral water and packaged drinking water. The transparency of such containers shall not be less than 85 percent in light transmittance.

Provided further that food grade packaging materials such as paper and paperboard, glass containers, metal or metal alloys, which may or may not contain plastic as component compatible with the water to be packaged may also be used. In such cases requirements of transparency would not apply.

Schedule–III Plastic Materials intended to come in contact with food products, pharmaceuticals and drinking water

S.No	Types of Plastic Material	Indian Standards specifications to be conformed
	Polyethylene in contact with foodstuffs	IS: 10146
	Styrene Polymers in contact with foodstuffs	IS: 10142
	Polyvinyl Chloride (PVC), in contact with foodstuffs	IS: 10151
	Polypropylene in contact with foodstuffs	IS: 10910
	Ionomer Resins in contact with foodstuffs	IS: 11434
	Ethylene Acrylic Acid (EAA) copolymer	IS: 11704

S.No	Types of Plastic Material	Indian Standards specifications to be conformed
	Poly alkyleneterephathalates (PET).	IS: 12252
	Nylon 6 Polymer	IS: 12247
	Ethylene Vinyl Acetate (EVA) copolymers	IS: 13601
	Ethylene Metha Acrylic Acid (EMAA) copolymers and terpolymers	IS: 13576
	Polycarbonate Resins	IS 14971
	Flexible Packaging Materials for packaging of Edible Oils, Ghee and Vanaspati	IS 14636
	Polyalkylene Terephthalates (PET & PBT) for Moulding and Extrusion	IS 13193
	Polyethylene Films and Sheets	IS 2508
	Linear Low Density Polyethylene (LLDPE) Films	IS 14500
	High Density Polyethylene Materials for Moulding and Extrusion	IS 7328
	Melamine-Formaldehyde Resins	IS 14999
	Low Density Polyethylene Films	IS 2508
	Blow Moulded Polyolefin Containers - Part 2 : Over 5 Litres, Up to and Including 60 Litres Capacity	IS 7408
	Stretch Cling Films	IS 14995

(b) All packaging materials of plastic origin shall pass the prescribed overall migration limit of 60mg/kg or 10mg/dm^2 when tested as per IS 9845 with no visible colour migration.

(c) Plastic materials and articles shall not release the substances in quantities exceeding the specific migration limits listed under Table 1.

Table 1: Requirement for specific migration limits of substances from plastic materials intended to be in contact with articles of food

S.No	Substance	Maximum Migration Limit (mg/Kg)
1	Barium	1.0
2	Cobalt	0.05
3	Copper	5.0
4	Iron	48.0
5	Lithium	0.6
6	Manganese	0.6
7	Zinc	25.0
8	Antimony	0.04
9	Phthalic acid, bis (2-ethylhexyl) ester (DEHP)	1.5

(d) Pigments or Colorants for use in plastics in contact with food products and drinking water shall conform to IS: 9833.

(e) Products made of recycled plastics including carry bags may be used for packaging, storing, carrying or dispensing of food products as and when standards and guidelines are framed by the Food Authority. Such packaging materials shall also comply with any other national standards/regulations as applicable

12.5.5. Packaging Material for Specific Food Products

A list of suggestive packaging materials which may be used for packaging of food products falling under the specified categories is provided in Schedule – IV:

Provided that this is an indicative list not restricting the use of any other packaging material complying with the specified standards.

Schedule – IV List of suggestive packaging materials

Si.No.	Product Category	Types of Packaging materials
1.	Milk and milk products	• Glass bottle with metal caps or plastic (polypropylene (PP) or High-densitypolyethylene (HDPE)) caps. • Rigid Plastic container made of PET with plastic (polypropylene (PP) or High-densitypolyethylene (HDPE)) caps. • Rigid Plastic container made up of High density polyethylene (HDPE) or Polypropylene (PP) or Polystyrene (PS) with Plastic (PP or High-densitypolyethylene (HDPE) caps. • Flexible plastic pouch made of polyethylene (PE) or Polypropylene (PP) based co-extruded multilayered material. • Aseptic and flexible packaging material (Paperboard or Aluminium foil or polyethylene) based multilayered structure. • Tinplate container. • Paper based lined cartons with liner made ofaluminium foil based laminated structure. • Plastic based polypropylene (PP) or polystyrene(PS) cups with paper or Peel-off lid. • Wax coated paper butter wrappers. • Paper and Paper Board based folding cartoninside butter wrapped with butter paper. • Metal Containers with plastic polypropylene (PP)caps or metal or plastic lid. • Plastic pet container with plastic lid. • Thermoform cup or tray with paper or peel-offlids. • Paper and Paper Board setup boxes with orwithout lamination –plastic film inside.

Si.No.	Product Category	Types of Packaging materials
		• Paper and Paper Board setup boxes with or • without grease proof paper placed inside. • Plastic Based multi layered flexible laminated heat sealed pouch. • Mud or clay pots. • Thermoformed Plastic container (blister pack) with aluminium foil or polyethylene (PE) based lid.
2.	Fats, oils and fat emulsions	• Tinplate container. • Glass bottle with metal caps or plastic polypropylene (PP) or High-density polyethylene (HDPE) caps. • Plastic rigid container (jar) made of High-density polyethylene (HDPE). • Plastic bottle or Jar Polyethylene terephthalate (PET) with plastic caps. • Plastic Pouch made of Multi layered laminated orco-extruded structure. • Aseptic and flexible packaging material (Paperboard or Aluminium foil or polyethylene) based multilayered structure. • Plastic laminated pouch in duplex board box (Bagin Box). • Thermoformed plastic based jar with plastic caps. • Paper based lined cartons with liner made of aluminium foil based laminated structure.
3.	Fruit & Vegetable products	• Glass bottle with metal caps or plastic (polypropylene (PP) or High-density polyethylene(HDPE) caps. • Aluminium can with easy open end. • Tinplate container. • Aseptic and flexible packaging material (Paperboard or Aluminium foil or polyethylene) based multilayered structure. • Plastic rigid container (jar) made of either High- density polyethylene (HDPE) or Co-extrudedstructure with Plastic (polypropylene (PP) or High-density polyethylene (HDPE) caps. • Stand up Pouch made up of Plastic based structure with plastic spout. • Flexible Plastic pouch made of either polyethylene (PE) or Laminated structure. • Thermoformed Plastic container (blister Pack) with aluminium foil or polyethylene (PE) based lid. • Plastic jar (Co-extruded) with metal caps. • Plastic trays with overwrap. • Polyethylene terephthalate (PET) or polypropylene (PP) or Poly Vinyl Chloride (PVC) Punnets

Si.No.	Product Category	Types of Packaging materials
4.	Sweets and Confectionery	• Metal Containers with plastic polypropylene (PP) caps or metal or plastic lid Plastic based multilayered laminated Heat sealed pouches. • Composite containers made up of Paper Board or Aluminium foil or plastic base films with plastic or metal lids. • Plastic based rigid containers. • Foil wrap. • Plastic film based twist wraps (Polyethylene terephthalate (PET) or polypropylene (PP) or PolyVinyl Chloride (PVC). • Thermoformed tray and punnet with lid. • Glass bottle with metal or plastic caps. • Plastic cups with film lid.
5.	Cereals and cereal products	• Tin container. • Aluminium Foil Based laminated pouch in metalcontainer. • Wrapper made of wax coated paper. • Wrapper made of three layered laminatedstructure. • Plastic based multilayered laminated pouch (heatsealed). • Plastic based thermoform container with plasticlid. • Lined carton with liner made of multi layeredlaminated structure. • Plastic based multilayered laminated structuredZipper pouch. • Thermoform trays with plastic lids or over wraps. • Glass bottle with metal caps. • Polyethylene terephthalate (PET) or Plastic basedrigid containers with metal or plastic (polypropylene (PP) or High-density polyethylene(HDPE) caps • Plastic films or co-extruded film or polypropylene (PP) or polyethylene (PE)
6.	Meat and Meat Productsor Poultry Products	• Glass jars with plastic (polypropylene (PP) orHigh-density polyethylene (HDPE) caps. • Metal Containers with metal lid (lacquered tincontainers). • Plastic based flexible pouches in paper & paperBoard carton. • Plastic based multilayered flexible laminates heatsealed pouches. • Plastic tray with overwrap. • Aluminium foil wrap. • Polyethylene terephthalate (PET) punnets orcontainers with plastic caps

Si.No.	Product Category	Types of Packaging materials
7.	Fish and fish products or Seafood	• Glass jars with plastic (PP or High-densitypolyethylene (HDPE) caps. • Metal Containers with metal lid (lacquered tincontainers). • Polyethylene terephthalate (PET) punnets orcontainers with plastic caps. • Plastic based multilayered flexible laminates heatsealed pouches. • Plastic tray with overwrap.
8.	Sweetening agentsincluding Honey	• Glass bottle with Metal Caps or Plastic (polypropylene (PP) or High-density polyethylene(HDPE) Caps. • Plastic based Thermoformed container. • Blister Pack with foil or polyethylene lid. • Polyethylene Terephthalate (PET) container withPlastic Caps. • Plastic laminated Tube.
9.	Salt, spices, Condimentsand related products	• Glass bottle with metal lid or plastic (polypropylene (PP) or High-density polyethylene(HDPE) caps. • Plastic based rigid container with Plastic cap (Polyethylene terephthalate (PET) and High- density polyethylene (HDPE) Containers). • Paper & Paper board or Aluminium foil or PlasticFilm based Composite Container. • Folding cartons with Plastic based flexible laminated structure (heat sealed) pouch placed inside. • Plastic based multilayered layered laminated pouch (heat sealed).
10.	Beverages (other thanDairy and Fruits & vegetables based)	• Plastic bottles made of either Polyethylene terephthalate (PET) or Polycarbonate (PC) with Plastic (Polypropylene (PP) or High-density polyethylene (HDPE) or Aluminium caps. • Heat sealed Plastic pouches made of Polyethylene(PE). • Glass bottles with metal caps or plastic caps. • Plastic pouches made up of Polyethylene (PE) in Corrugated fibre board Boxes. • Aluminium can with easy open end. • Tinplate container. • Plastic pouch made of laminated structure. • Aseptic and flexible packaging material (Paperboard or Aluminium foil or polyethylene) based multilayered structure. • Plastic based multi layered structure heat sealed pouches. • Plastic based multi layered structure heat sealed Zipper pouches or stand up pouches. • Metal Containers with plastic or Polypropylene (PP) caps or metal or plastic lid, Rigid Plastic container with plastic caps (Polypropylene (PP) Caps). • Wooden cask (for wines).

References

Agriculture Produce Grading and Marking (AGMARK) Act, 1937 https://legislative.gov.in/sites/default/files/A1937-1.pdf

Bureau of Indian Standards (BIS) Act, 1986 and BIS rules, 1987 https://www.bis.gov.in/bis-act-rules-regulations/118330-2/

Dawson, R. J. (1995). The role of the Codex Alimentarius Commission in setting food standards and the SPS agreement implementation. Food Control, 6(5), 261-265.

Food Safety and Standards (Packaging) Regulation, 2018 FSSAI https://www.fssai.gov.in/upload/uploadfiles/files/Gazette_Notification_Packaging_03_01_2019.pdf

Heckman, J. H. (2005). Food packaging regulation in the United States and the European Union. Regulatory Toxicology and Pharmacology, 42(1), 96-122.

Joint FAO/WHO Codex Alimentarius Commission. (1992). Codex alimentarius. Food & Agriculture Org..

Premavalli, K. S. (2013). Food Laws and Regulations. Food Science, 365.

Sarkar, S., & Aparna, K. (2020). Food packaging and storage. Research Trends in Home Science and Extension AkiNik Pub, 3, 27-51.

The Standards of Weights and Measures Act, 1976 https://consumeraffairs.nic.in/en/more/archive/the-standards-of-weights-and-measures-act

Trachtman, J. P. (2002). The Technical Barriers to Trade Agreement, the Sanitary and Phytosanitary Measures Agreement, and the General Agreement on Tariffs and Trade< i> A Map of the World Trade Organization Law of Domestic Re. Journal of World Trade, 36(5).

Veraart, R., & Rijk, R. (Eds.). (2010). Global legislation for food packaging materials. John Wiley & Sons.

13

Labelling in Packaging

Ajay Patel[1], Priyanka Sharma[2], S.N. Naik[3], Kshitiz Kumar[4]

[1]*Department of Agriculture, SRM University Delhi-NCR, Sonepat Haryana, India*
[2]*Centre for Early Childhood Development and Research, Jamia Millia Islamia India*
[3]*Center for Rural Development and Technology, Indian Institute of Technology Delhi, India*
[4]*A.D. Patel Institute of Technology, CVM University, V V Nagar, Gujarat, India*

13.1. Introduction

Any written, printed, or graphic matter that is present on the label and serves the purpose of promoting the sale of any product is known as food label (WHO). It is believed in general that food packaging and proper labelling plays an important role in food safety as it gives information regarding allergens present, manufacturing site, expiry date, nutritional information etc.

Consumption of unhealthy diet leads to many non-communicable diseases. Therefore, it is the responsibilities of food manufacturers and regulating bodies such as FSSAI and BIS to work hand in hand to provide safe and healthy food to the consumers. Labelling is an important tool useful to disseminate nutritional information to the consumers. Labelling promotes healthy eating among consumers and gives them an option at the time of purchase to choose healthy food. Information in labelling the enables consumers to choose appropriate food according to their diets and lifestyle (Grunert and Wills, 2007). In addition, the information related to nutritional content of the food must be present in such a way that it well understood by every consumer (Wills et al., 2009).

In old times, labelling of any particular food was limited to name, price, quantity, identity of the manufacturer. Now the scope of the food labelling has changed and it is one of the prime factors in increasing the market value and sale of any product. Food label also helps to bridge the gap between consumers and the food ingredients. Nowadays, consumers are smart and they need detailed and well informed labels that has become an indispensable part of today‘s consumption scenario (Singla, 2010). Principal reason for nutrition labelling is that the consumers have a right to know what is in the purchased food so that consumers could take better decisions while buying food items (Rotfeld and Taylor, 2009). This chapter contains the labelling requirements covered under Food Safety and Standards (Labelling and Display) Regulations, 2020 adopted by FSSAI (Food Safety and Standards Authority of India).

13.2. General Requirements of Labelling

All the pre-packaged food items should carry a label which has all the information about the food inside the package. The language shall be given either in Hindi or English. The use of any other language is prohibited.

The label in the packaging material should be attached in such a way that they do not separate from packaging materials easily. The labelling should not contain misleading, false and deceptive information on the packaging of pre-packaged food. The label should be legible, clear, indelible while purchase and use by the consumer. If the food is packed in a wrapper, the wrapper shall bear all the necessary information on the label.

13.3. Labelling of Pre-packaged Food Items

In addition to the above mentioned general requirements, every package of food should contain the following information on the label:

(i) Name of the food item: Trade name or description of the food item contained in the package should be mentioned in the package

(ii) Details of ingredients: Except where food contains a single ingredient, a list of ingredients should be declared on the label. A suitable title such as "Ingredients" or "Composition"should be used to mention the complete list of ingredients. The name of each ingredient should be arranged in descending order of their composition present in the food item at the time of their manufacturing (by weight or volume). The name of the ingredients in the ingredients list shall be specific. Different 'Class Titles' may be used on the labels. Different class of food item and their class titles are mentioned in the Table 1:

Table 1: Details of classes of food items and their respective class titles

S.No.	Classes	Class Titles
1	Edible vegetable oils	Mention edible oil with specific name such as soyabean oil, sunflower oil, groundnut oil, rice bran oil etc.
2	Edible vegetable fat	Mention vegetable fat type such as vegetable fat, hydrogenated or partially hydrogenated oils, edible vegetable fats, margarine and fat spreads.
3	Animal fat or oil except milk fat	Source of fat/oil shall be mentioned such as beef fat, pork fat etc.
4	Starches: except chemically modified starches	Starch
5	All species of fish where the fish constitutes an ingredient of another food and provided that the labelling and presentation of such food does not refer to a species of fish	Fish
6	All types of meat where such meat constitutes an ingredient of another food and provided that the labelling and presentation of such a food does not refer to a specific type of meat	Meat
7	All types of cheese where cheese or mixture of cheeses constitutes an ingredient of another food and provided that the labelling and presentation of such food does not refer to a specific type of cheese	Cheese
8	All spices, herbs and condiments and their extracts	Spices, herbs and condiments or mixed spices/ condiments as appropriate
9	All types of gum or preparations used in the manufacture of gum base for chewing gum	Gum Base
10	Anhydrous dextrose and dextrose monohydrate	Dextrose or Glucose
11	Sucrose	Sugar
12	All types of Caseinates	Caseinates
13	Press, expeller or refined cocoa butter	Cocoa butter
14	All crystallized fruit or vegetable	Crystallized fruit/vegetable
15	All milk and milk products derived solely from milk	Milk solids
16	Cocoa bean, Cocoa nib, Cocoa mass, Cocoa press cakes, Cocoa powder (Fine/ Dust)	Cocoa solids
17	All vitamin(s) and their compounds	Vitamin(s)
18	All minerals and trace elements and their compounds/salts	Mineral (s) and/or trace element(s)

Specific names shall be mentioned on the labelling for lard fat, pork fat and beef fat/extract. Added water shall be declared in the list of ingredients except in cases where water forms part of an ingredient, such as, brine, syrup or broth, used in the compound food and so declared in the list of ingredients. If water or other volatile ingredients gets evaporated in the course of manufacture then they need not be declared. In case of concentrated/condensed/dehydrated foods, statement such as "***Ingredients of the product when prepared in accordance with the directions on the label***" should be given.

In a condition where compounds ingredients (ingredients itself formed by combination of two or more ingredients) are present, the ingredients of compound ingredients shall be mentioned in a bracket in decreasing order of their weight and volume. If compound ingredient, constitutes less than 5% of the food , then ingredient of compound ingredients should not be given in the list of ingredients. However, syrup, brine, or broth, added water shall always be mentioned on the label under list of ingredients.

The list of ingredients shall be given in descending order of their weight and volume. If the food is sold as mixture or combination then the percentage of the ingredients used at the time of manufacture of food should be mentioned. In case of flavoring ingredients, it is not mandatory to mention in ingredient list. Also if drained net weight is indicated on label or pictorial representation of serving suggestion is made, disclosure of ingredients is not required.

The declaration with regard to addition of fruit pulp in ready to serve beverages shall be mentioned on the label. Similarly, bottle containing liquid milk as an ingredient shall also be given.

13.3.1. Nutritional Information

Information related to nutrition is to be provided in per 100 gm or per 100 ml or per single consumption pack of the product and per serve percentage (%) contribution to Recommended Dietary Allowance calculated on the basis of 2000kcal energy, 67 g total fat, 22 g saturated fat, 2 g trans fat, 50 g added sugar and 2000 mg of sodium (5 g salt) requirement for average adult per day. The amount of different ingredients such as protein content, fat content, carbohydrate (specific quantity of sugar) shall be mentioned in gram, while the energy value in kcal. The amount of any another ingredient if promoted as health benefits shall also be provided.

Type and amount of fatty acids, amount of cholesterol in milligrams, amount of saturated fatty acids, monounsaturated fatty acids (MUFA) and polyunsaturated fatty acids (PUFA) in grams, trans fatty acids in grams shall be mentioned on the label. Vitamins and minerals are to be given in metric units.

In case of enriched food with proteins, minerals, vitamins, enzymes or amino acids etc., the amount of added nutrients on the label shall be mentioned. For raw agricultural items such as spices, wheat, rice, herbs, jaggery, sugar, table salt, non-nutritive items like packaged drinking water, alcoholic beverages, tea, coffee, fruits and vegetables, minimal processed/prepackaged assorted vegetables, pickles, papad, or foods served for immediate consumption the nutritional information may not be necessary.

Food material containing hydrogenated vegetable fat or bakery shortening should contain a label 'hydrogenated vegetable fats or bakery shortening used-contains trans fats'.

For the food items where the level of trans fat is less than 0.2 grams per serving of food, a health claim called ***TRANS FAT FREE*** can be made. Similarly, for products where the saturated fat does not exceed 0.1 gm per 100 gm or 100 ml of food, health claim such as ***SATURATED FAT FREE*** can be made.

13.3.2. Declaration Regarding Vegetarian and Non-vegetarian Logo

The information regarding vegetarian and non-vegetarian types of food product is to be declared on label through symbol and colour code as depicted in the Figure :

Fig. 1: Symbol for vegetarian food

Fig. 2. Symbol for non-vegetarian food

The symbol for vegetarian food consist of a green colour filled circle inside a square with green outline while symbol for non-vegeterian food consist of a brown colour filled triangle inside a square with brown outline having the sides not less than the minimum size specified in the Table mentioned.

Table 2: Size of the logo

S.No.	Area of principal display panel (cm^2)	Minimum size of diameters of circle (mm)	Minimum size of each side of triangle (mm)	Minimum size of each side of square (mm)
1	Upto 100	3	2.5	6
2	Above 100 - 500 cm^2	4	3.5	8
3	Above 500 – 2500	6	5	12
4	Above 2500	8	7	16

The vegetarian and non-vegetarian logo should be printed on the labels, containers, pamphlets, leaflets, advertisements given in the mass media. If article of food contains egg only as non-vegetarian ingredient, the manufacturer, or packer or seller may give declaration to this effect in addition to the said symbol. In case of mineral water, packaged drinking water, carbonated water, alcoholic drinks, liquid milk, milk powders, such logo shall not be applicable.

13.3.3. Labelling Regarding Food Additives

In case of usage of food additives, the class titles shall be used together with the specific names or by the international numbering system for food additives (INS), which is a Europe based food additive system. Its function is to give short names for the individual additives. The few examples of food additive along with their number, name, class and uses are given in table:

Table 3: Food additive along with their number, name, class and uses

Number	Name	Class	Uses
E150	Caramel	Coloring agent	Used in biscuits, malt vinegar, marmalade, soft drinks and beef products.
E160b	Annatto	Coloring agent	It gives creamy color to pastry, coffee, cheese and its products
E160d	Lycopene	Coloring agent	It uses in dairy and dairy products and gives color yellow to red
E211	Sodium Benzoate	Preservative	In different bottled beverages
E220	Sulfur Dioxide	Preservative	Uses in mixed peel and meat products.
E234	Nisin	Preservative	Used in dairy and meat product
E341	Calcium Phosphate	Rising agent.	Baked products.
E401	Sodium Alginate	Thickener	In gem and gelly
E621	Monosodium Glutamate (MSG	Flavour Enhancer	In snacks
E951	Aspartame	Sweeteners	Confectionary
E558	Bentonite	Anti-caking agent	

In case of a product containing extraneous coloring matter, following declaration to be given under the list of the ingredients.

Contains Permitted Natural Colour(S) Or Contains Permitted Synthetic Food Colour(S) or Contains Permitted Natural And Synthetic Food Colour(S)

The flavoring agents should also be mentioned on the label with the statement such as

Contains Added Flavour

In case of combined declaration of colors and flavours, the International Numerical Identification number of colors used shall also be indicated either under the list of ingredients. Label should also bear the percentage of dye content present in the foods.

Mandatory Declarations regarding food additives

Every package of food containing the following ingredients/additives should bear the following declarations on the label in a rectangular box, namely

Table 4: Mandatory Declarations regarding food additives on the labeling

S.No	Ingredients/additives	Declarations on the labelling
1	10% or more polyols	Polyols may have laxative effect
2	10% or more polydextrose	Polydextrose may have laxative effect
3	Added caffeine	CONTAINS CAFFEINE Provided if caffeine is added in the products, the quantity of the added caffeine (in ppm) shall also be declared in the list of ingredients.
4	Isomaltulose	Contains Isomaltulose............ (kcal) per 100gm or 100ml
5	10 per cent. or more Sorbitol and Sorbitol syrup	May have laxative effect, cause bloating and diarrhea in children; and reduce calcium absorption in post-menopausal women

Every package containing the following articles of food shall bear the following declarations on the label in a rectangular box, namely, -

Table 5: Mandatory Declarations regarding food article

S.No	Articles of food	Declarations
1	Maida treated with improver or bleaching agents	Refinedwheat flour (maida)treated with improver/ bleaching agents, to be used by bakeries only
2	Dried Glucose Syrup containing sulphur dioxide exceeding 40 ppm	Dried Glucose Syrup For Use In Sugar Confectionery Only
3	Fruit squash by whatever name it is sold, containing additional sodium or potassium salt	Contains additional sodium/potassium salt

S.No	Articles of food	Declarations
4	Flavour emulsion and flavour paste meant for use in carbonated or non-carbonated beverages	Flavour emulsion and flavour paste for use in carbonated or non-carbonated beverages only
5	Cheese(s), if coated/packed in food grade waxes	Coated wax to be removed before consumption
6	Frozen Desert/Frozen Confection	Frozen Desserts/Frozen Confection is made with _______________Edible Vegetable Oil*/and Vegetable Fat*
7	Common Salt	Common Salt For Iodisation*/ Iron Fortification*/ Animal Use*/ Preservation Not Meant For Direct Consumption*/Infant Food Products*/Medicine*/ Industrial Use*
8	Fresh fruit if coated with wax	Coated with Wax (give name of wax)
9	Gelatin meant for human consumption	Gelatin Food Grade
10	Aspartame (Methyl ester), Acesulfame Potassium, Aspartame-Acesulfame salt, Sucralose, SACCHARINS, Neotame, Steviol Glycoside and Polyols marketed as "Table Top Sweetener"	i) Contains........... (Name of sweetener with purity and weight percent of marker compound) (ii) Not recommended for phenylketonurics; for children suffering from seizure disorders; pregnant and lactating mothers (in case of Aspartame (Methyl ester)) (iii) Not recommended for children; pregnant and lactating mothers (in case of Acesulfame Potassium) (iv)Not recommended for phenylketonurics; for children; pregnant and lactating mothers (in case off Aspartame-Acesulfame salt) (v) Not recommended for children" (in case of Saccharins) (vi) Polyols may have laxative effect (in case of Polyols) (vii) May have laxative effect, cause bloating and diarrhea in children; and reduce calcium absorption in post-menopausal women (in case of Sorbitol and Sorbitol syrup).]

S.No	Articles of food	Declarations
11	Sweeteners mentioned under Appendix A of Food Safety and standards (Food Products standards and Food Additive) Regulations, 2011	i) This contains............................ (name of the sweetener). (ii)*Not recommended for phenylketonurics; for children suffering from seizure disorders; pregnant and lactating mothers (if Aspartame is added) (iii)*Not recommended for children; pregnant and lactating mothers" (if Acesulfame potassium is added) (iv)*Not recommended for phenylketonurics; for children; pregnant and lactating mothers (if Aspartame-Acesulfame salt is added) (v) *Not recommended for children (if SACCHARINS is added)
	Mixture of Sweeteners mentioned under Appendix A of Food Safety and Standards (Food Products Standards and Food Additive) Regulation,2011	This .... (name of food) contains an admixture of .... (name of the sweeteners). Provided that in addition to the above declaration every package of food containing mixture of sweeteners shall declare the labelling requirement prescribed under these regulations for the individual sweeteners present in the mixture;
12		Note: In case of food package having surface area upto 100cm2 , the size of numerals & letters for such declarations shall not be less than 1.5 mm
13	Monosodium Glutamate	This package of (name of the food) .... contains added Monosodium Glutamate not Recommended for Infants Below -12 Months And Pregnant Wome

Every container of refined vegetable oil shall bear the following label, namely

Refined (name of the Oil) Oil

One time usable plastic bottles of packaged drinking water and mineral water shall carry the following declaration:

Crush the Bottle after use

or

Dispose the Bottle Responsibly

Every package of fat spread, milk products, milk based fruit drink, fermented milk products, soy and rice drink, cheese products, yoghurt products, spice sauces, salad dressings, juices and nectars containing added plant stanol esters shall bear the following label, namely

- *Contains Plant Stanol Esters (as Plant Stanols)---gm/100 gm or 100ml.*
- *Patients on cholesterol lowering medication should use the product under medicalsupervision.*
- *May not be nutritionally appropriate for pregnant and lactating women and childrenunder the age of five years.*
- Consumption of More Than 3G. Per Day, Total of Sterol, Stanols or Combination Therefof Shall be Avoided.

Every package of biscuits, bread, cakes, breakfast cereals, carbonated water, thermally processed fruits, fruit juices, fruit nectars, fruit beverages, fruit squashes, jam, jelly, fruit cheese, marmalade, dairy based drinks, milk powder, carbohydrate- based and milk product based sweets like gulabjamun, rosogolla, peda, khoya burfi, macroni products, noodles, pasta, sweets and confectionery, candies and icings, savories and snacks wherever the trehalose is added shall bear the following label, namely:—

Contains Trehalose

Every package of flakes and ready to eat dry breakfast cereals, noodles, pasta, salad dressings or toppings and spreads; table top fibre as filler or carrier, cereals and other snack food or savouries and bakery products including biscuits, cookies, bread, cake mix and pastries and other products where dextrin is allowed under Food Safety and standards (Food Products standards and Food Additive) Regulations, 2011, containing added Dietary Fibre (Dextrin-soluble fibre), shall bear the following declarations, namely:-

Contains Dietary Fibre (Dextrin) ----- (Source of soluble Dietary Fibre)

Fat spread, milk products, milk based fruit drink, fermented milk products, soy and rice drinks, cheese products, yoghurt products, spice sauces, salad dressings, juices and nectars, edible oils, and bakery products containing added plant sterol shall contain the following declarations, namely: -

- *Contains Plant Sterol (as Phytosterols) ---g/100g or 100ml.*
- Patients on cholesterol lowering medication should use the product under medical supervision
- May not be nutritionally appropriate for pregnant and lactating women and children under the age of five years.
- Consumption of More Than 3G/Day, Total of Sterol, Stanols, or Combination Thereof, Should Be Avoided

The label of a food, which has been treated with ionizing radiation, provided under regulation 2.13 of Food Safety and Standards (Food Products Standards and Food Additives) Regulations, 2011, shall bear the Radura logo (Figure) shall carry a written statement indicating the treatment in close proximity to the name of the food.

Processed By Radiation

Fig. 3: Radura logo for irradiated food

13.3.4. Details of Name and Complete Address of the Manufacturer

Complete name and address of marketing unit as well as manufacturing company shall be mentioned on the label. The name and address shall be preceded by the qualifying words "Manufactured by (Mfg by, Mfd by)" or "Marketed by (Mkt by)" or "Manufactured & Marketed by" or "Packed & Marketed by" as the case may be. In case, manufacturing company is different than marketing company than name and address of both the company shall be mentioned separately on the label. If an article of food is imported into India, the package of food shall also carry the name and complete address of the importer. If any food article manufactured outside India is packaged or bottled in India, the package containing such food article shall also bear on the label, the name of the country of origin of the food article and the name and complete address of the importer and the premises of packing or bottling unit in India.

13.3.5. Declaration of Net Quantity Retail Sale Price and Consumer Care on the Label

Net quantity, retail sale price and consumer care at the time of packaging of food articles shall be mentioned on label of every package as per provisions in Legal Metrology Act, 2009 (1 of 2010) and the Rules made there under.

The following commodities shall be packed in such quantities by weight, measure or number as are specified in the corresponding entries against them.

Table 6: Commodities to be packed in specified quantities

S.No	Commodities	Quantities in which to be packed
1	Baby Food	100g, 200g, 300g, 400g, 500g, 600,g, 700 g, 800 g, 900g, 1 kg, 2kg, 5 kg and 10 kg
2	Weaning food.	25g, 50g, 75g, 100g, 150g, 200g, 250g, 300g and thereafter in multiples of 100g up to 1 kg.
3	Bread including brown bread but excluding bun.	100g and there after in of multiples 100g.
4	Un-canned packages of butter and margarine	25 g, 50 g, 100 g, 200 g, 500 g, 1 kg, 2 kg, 5 kg, and thereafter in multiples of 5 kg.
5	Cereals and Pulses	100g, 200g, 500g, 1 kg, 2 kg, 5 kg and thereafter multiples of 5 kg
6	Coffee	25g, 50g, 100g, 200g, 250g, 500g, 1kg and thereafter in multiples of 1kg.
7	Tea	25g, 50g, 100g, 125g 250g, 500g, 1kg and thereafter in multiples of 1kg.
8	Materials which may be constituted or reconstituted as beverages	25 g, 50 g, 100 g, 200 g, 500 g, 1 kg and thereafter in multiples of 1 kg.
9	Edible Oils Vanaspati, ghee, butter oil	50 g, 100 g, 200 g, 500 g, 1 kg, 2 kg, 3 kg, 5 kg and thereafter in multiples of 5 kg If net quantity is declared by volume the same number in milliliters or liters, as the case may be. If the net quantity is declared by volume, then the equivalent quantity in terms of mass to be declared in brackets in same size of letters/ numerals
10	Milk Powder.	Below 50g no restriction, 50 g, 100g, 200g, 500 g, 1 kg and thereafter in multiples of 500 g.
11	Rice (powdered), flour, atta, rawa and suji	100g, 200g, 500g, 1kg, 2kg, 5 kg and thereafter in multiples of 5 kg.
12	Salt	Below 50g in multiples of 10g, 50g, 100g, 200g, 500g, 750g, 1 kg, 2 kg, 5 kg and thereafter in multiples of 5 kg
13	Aerated soft drinks, non-alcoholic beverages.	65 ml (fruit based drinks only), 100 ml, 125ml(fruit based drinks only), 150 ml, 200 ml, 250 ml, 300 ml, 330ml(in cans only), 500 ml, 750 ml, 1 litre, 1.5 litre, 2 litre, 3 litre, 4 litre and 5 litre.
14	Mineral water and drinking water	1 kg, 2 kg, 5 kg, 10 kg, 20 kg, 25 kg, 40 kg (for White cement only) and 50 kg.

13.3.6. Lot or Code or Batch identification

It is very important in the process of food recall. A food is recalled with the help of lot/batch number. The label should clearly display the code number or batch number.

13.3.7. Date of Manufacture or Packing

The date of manufacturing is a mandatory feature to be mentioned on the label. In case a food has a shelf life more than three months, the manufacturing month and year is to be given on the label. Whereas if a food article is having a shelf life of less than 3 months, manufacturing date, month and year is to be given on the label.

13.3.8. Declaration of Best Before and Use by Date

"Date of manufacture or packaging" and "Expiry/Use by" shall be declared on the label. However, expression "Best before" may also be used as optional or additional information. The day, month and year using the DD/MM/YY format for products with a short shelf life of up to 3 months; the month and the year for products with a shelf life of more than three months, shall be declared in un-coded numerical sequence except that the month shall be indicated by capital letters and abbreviations (at least first three letters of the month) may be used

"Expiry" declaration is not required in case of Fresh fruits and vegetables, including potatoes which have not been peeled, cut or similarly treated, wines, liquors, and other alcoholic beverages where the alcohol content is more than 10% by volume, vinegar, Sugar boiled confectionery, Food grade salt for industrial use, solid sugars, chewing gum and bubble gum. It is mandatory to mention best before or use by date or expiry date in case of milk and milk products such as curd, infant food and infant milk substitute.

13.3.9. Instruction for use Declaration

Either in pictorial form or words or through flow chart the instruction to use along with the quantity of additional things such as water, salt etc. shall be mentioned on the label.

13.3.10. General Conditions Regarding Declaration on the Label

It is mandatory to provide any declaration, pictorial, written, printed or graphic matter on the label. A label shall be legible, definite and unambiguous in nature. It is to be noted that all declarations with regard to food item shall be displayed on the wrapper or carton. Any misleading claim, statement, fancy names, names like pure on the label shall be avoided.

13.3.11. FSSAI logo and license number

The package label should display FSSAI logo and license number in contrast color to the background as below.

Lic. No. XXXXXXXXXXXXXX

Fig. 4: FSSAI Logo and License number to be displayed on Label

If the license number of the manufacturer or marketer or packer or bottler, as the case may be, is different from the brand owner then their logo shall also be displayed on the label. In case of imported food products, the importer shall display FSSAI logo and license number along with name and address of importer.

13.3.12. Labelling of Imported Foods

Labelling requirements for imported products shall be governed by the Food Safety and Standards (Import) Regulations, 2017 in addition to the requirement mentioned in these regulations.

13.3.13. Country of Origin for Imported Foods

(i) The country of origin of the food shall be declared on the label of food imported into India.

(ii) When a food undergoes processing in a second country which changes its nature, the country in which the processing is performed resulting in change in HS Code at the 6 digit level shall be considered to be the county of origin for the purposes of labelling.

13.3.14. Declaration Regarding Food Allergen

The following foods and ingredients which are known to cause allergy shall be declared separately as Contains............................ (Name of allergy causing ingredients)

(i) Cereals containing gluten; i.e., wheat, rye, barley, oats, spelt or their hybridized strains and products of these

(ii) Crustacean and their products (To be declared as Crustacean);

(iii) Milk & Milk products(To be declared as Milk);

(iv) Eggs and egg products (To be declared as Egg);

(v) Fish and fish products (To be declared as Fish);

(vi) Peanuts, tree nuts (e.g. almonds, walnuts, pistachio, cashew nuts) and their products (To be declared as Nut);

(vii) Soybeans and their products (To be declared as Soy);

(viii) Sulphite in concentrations of 10mg/kg or more (To be declared as sulphite)

In case presence of ingredients due to cross contamination which are known to cause allergy may be declared separately as May Contains......................... (Name of allergy causing ingredients).

Such declaration is not required in case of oils derived from these ingredients." Raw agricultural commodities are exempted from the allergen labelling requirements.

13.3.15. Food not for Human Consumption

Every package of food material sold in retail but which is not meant for human consumption example Pooja water, Ghee for diya, Oil for Pooja etc. shall bear a declaration to this effect by a symbol as stipulated below. The symbol shall consist of a black colour cross inside a square.

Fig. 5: Symbol for food not for human consumption

13.4. Principal Display Panel

The information on principal display panel of any food item shall be grouped together. The area of principle display panel should be according to conditions given below:

(i) **Rectangular container**: The principal display panel area should be at least 40% of the resulted area which is resulted after multiplying the height with width of the containers.

(ii) **Cylindrical, oval, round, near oval, near cylindrical containers**: The area which is obtained by multiplying the height with the average circumference of the containers. The area must be 20 % of this obtained area for principal display panel.

c. **Container with random shape**: In that case the total surface area of the package/container must be calculated and the area for the principal display panel must be 20% of the total surface area of the package.

If a label has been affixed, in that case principal display panel must be must not less than 10% of the total surface area of the package/container.

13.5. Restrictions on Labelling

Under Food Safety and Standards (Packaging and Labelling) Regulations, 2011, there have been certain restrictions provided for labelling of different categories of food items.

13.5.1. Restrictions on Infant Milk Substitute and Infant Food

In case of infant milk substitute products, where the standards of the products are not notified by the food authority, in such case, approval of such articles to be taken for the product as well as label before the product is released in the market. The words like "IMPORTANT NOTICE" in capital letters to be provided in every container of infant milk substitute or any label of such product. The information to be provided in a clear and easy to read manner. Various conditions are given as follows:

(i) A statement such as "***Mother's milk is best for your baby***" in the capital letters shall be displayed in the central panel of every container/package. The color of the above statement shall be different form the color of label.

(ii) On every article of infant milk substitute, ***"infant food shall be given only after the age of six months and upto the age of two years"***to be mentioned.

(iii) "***Infant food or infant food substitute should be used only after the consultation of health workers***" shall be given as a statement on the label.

(iv) Directions for its usage shall be provided on the label.

(v) Following warning statement is to be given on the label:

- ***Infant milk substitute or infant food cannot be only source of nourishment for the infant.***

- ***Careful and hygienic preparation of infant foods or infant milk substitute is most essential for health.***

(vi) The energy value in kcal or Joules, approximate composition of nutrients per 100 gm of the product shall be given on the label.

(vii) The storage conditions are to be mentioned on the label.

(viii) Scoop size and the instruction for correct measuring should be given.

(ix) Batch No. and days, Month and Year of its manufacture and expiry date shall be given on the label.

(x) Food additives information, if added, is required to be mentioned on the label.

The photograph of baby or women or both shall not be present on the label of containers of infant milk substitute and infant milk. "Full Protein Food", "Energy Food", "Complete food" or "Health Food" or any other similar statement shall not be mentioned on the label.

The Container of Infant Milk Substitute

It is mandatory to write premature baby (born before 37 weeks or low weight at the time of birth (less than 2.5 kg) on the food items specifically manufactured for premature baby (born before 37 weeks or low weight at the time of birth (less than 2.5 kg) on the label in capital letters along with its product name. If mother's milk is available, it is advised to withdraw infant milk substitute. A statement ***"To be taken under medical advice"*** is required to be mentioned on the label.

"LACTOSE-FREE or SUCROSE-FREE or LACTOSE and SUCROSE-FREE" shall be mentioned on the container for lactose or lactose and sucrose intolerant infants. A declarative statement "HYPOALLERGENIC FORMULA" in capital letters to be mentioned on the container of infant milk substitute meant for infants with allergy to cow's /buffalo's milk protein or soy protein or label affixed thereto.

Labelling of Edible Oils and Fats

Expressions like "Super-Refined", "Extra-Refined", "Micro-Refined", "Double-Refined", Ultra-Refined", "Anti-Cholesterol", "Cholesterol Fighter", "Soothing to Heart", "Cholesterol Friendly", "Saturated Fat Free" or such other expressions shall not be used in the labelling of fats and oils.

It is mandatory to give Indian Standards Institution (ISI) certification mark in the solvent container. The name or trade name of the solvent extracted oil or de oiled meal or edible flour is to be mentioned.

A declaration such as "free from Argemone Oil" should be provided on the label in the container of vanaspati, margarine, bakery shortening, blended vegetable oils, mixed fat spread and refined vegetable oil. The mass and volume of the contents should also be given. On every container, a statement **Not To Be Sold Loose** in bold capital letters along with the name of the product shall be given on the central panel. Name of the oil is to be given on every container of refined vegetable oil. The container of imported edible oil shall also bear the word imported as prefix.

Every package of vanaspati made from more than 30 percent of Rice bran oil shall bear the following label, namely: -

'This package of vanaspati is made from more than 30 per cent Rice bran oil by weight"

In case rice bran oil which is physically refined is used as one of the ingredients in Vanaspati, it shall be declared in the ingredient list on the label as 'Physically Refined Rice Bran Oil'

Every package containing an admixture of edible oils shall carry the following label declaration in bold capital letter immediately below its brand name or trade name on the front of pack, namely: -

"**Multi-Source Edible Oil**"

(Name and nature* of edible oil) ....per cent. by weight

(Name and nature* of edible oil) ....per cent. by weight

(*i.e. in raw or refined form)

For pack size less than one litre, the font size of the label declaration "MULTI-SOURCE EDIBLE OIL", shall not be less than 3 mm with the length of declaration statement as 35 mm minimum and for label declaration "Name and Nature of edible oil......per cent. by weight", font size shall not be less than 2 mm.

For pack size one litre to below 5 litre, the font size of the label declaration "Multi-Source Edible Oil" shall not be less than 4 mm with the length of declaration statement as 45 mm minimum and for label declaration "Name and Nature of edible oil......per cent. by weight" font size shall not be less than 2.5 mm.

For pack size five litre and above, the font size of the label declaration "Multi-Source Edible Oil" shall not be less than 10 mm and for label declaration "Name and Nature of edible oil......per cent. by weight" font size shall not be less than 3 mm.

Labelling of permitted food colors

The label should carry the words "Food colors" along with the chemical or commercial name. A mixture of permitted synthetic food colors for use in or upon food should carry the words "Food color Mixture" along with the chemical and the common or commercial name and color index of the dye-stuff. The ingredients used in the preparation of food color should also be mentioned on the label along with the words "Food color Preparation".

Labelling Prohibitions for Drinking Water (both packaged and mineral)

(i) No claims concerning medicinal (preventative, alleviative or curative) effects shall be made in respect of the properties of the product covered by the standard. Claims of other beneficial effects related to the health of the consumer shall not be made.

(ii) The name of the locality, hamlet or specified place may not form part of the trade name unless it refers to packaged water collected at the place designated by that trade name.

(iii) The use of any statement or of any pictorial device which may create confusion in the mind of the public or in any way mislead the public about the nature, origin, composition, and properties of such waters put on sale is prohibited.

13.6. Specific Labelling Requirements of Other Products

13.6.1. Coffee-Chicory Mixture

Every package containing a mixture of coffee and chicory shall have affixed to it a label upon which shall be printed the following declaration: Coffee blended with Chicory . This mixture contains

Coffee----percent

Chicory---percent

13.6.2. Concentrated Milk or Desiccated/Dried Milk

A container of Concentrated milk or Desiccated/Dried milk shall contain following details on the label:

Table 7: Information on the label of container of concentrated milk or Dried milk

S.No	Food article	Information on the label
1	Concentrated milk unsweetened	Equivalent to X--- liters of toned milk
2	Concentrated milk sweetened	Equivalent to X--- liters of toned milk with added sugar in gm
3	Concentrated skim milk unsweetened	Equivalent to X--- liters of skimmed milk
4	Concentrated skim milk sweetened	Equivalent to X--- liters of skimmed milk with added sugar in gm
5	For UHT milk	Sterilized by UHT process
6	Milk powder	Equivalent to X--- liters of toned milk
7	Milk powder with added lecithin	This container contains lecithin
8	Partly skimmed milk powder	Equivalent to X--- liters of partly skimmed milk having X--- percent fat
9	Skimmed milk powder	Equivalent to X--- liters of skimmed milk

13.6.3. Fluid Milk

The following alphabet letters should be mentioned on the caps of different categories of milk:

Table 8: Alphabet letters should be mentioned on the caps of different categories of milk

Milk type	**Letters denoted with**
Buffalo	B
Cow	C
Goat	G
Standardized	S
Toned	T
Doubled toned	DT
Skimmed milk	K
Pasteurized milk	P

13.6.4. Labelling of 'Gluten Free' Products

The term "Gluten Free" shall be printed in the immediate proximity of the name of the product in the case of products described in regulation 2.14 of the Food Safety and Standards (Food Products Standards and Food Additives) Regulations, 2011, namely: "Gluten Free"

In case any gluten free product is manufactured in a plant where gluten containing products are also manufactured this shall be declared on the label as "Processed in a plant where gluten containing products are manufactured"

13.6.5. Miscellaneous Foods

- If the packet of chillies have added edible oil in it then the label shall carry Chillies In This Package Contains An Admixture Of Not More Than 2 Percent Of..................(Name Of Oil) Edible Oil.
- If maida is treated with improver or bleaching agents then the label must carry a declaration Wheat Flour Treated With Improver/Bleaching Agents, To Be Used By Bakeries Only.
- It is mandatory to mention Iron Fortified Common Salt statement on a package of iron fortified common salt.
- "FLAVOURED TEA" Registration number is to be mentioned on the package of tea with added flavor.

13.7. Specific restrictions on Product labels

- The statement such as recommended by medical professional or approved by medical practitioners shall not be present on the labels of the food article.
- The word imitation shall not be written in the label or statement on any package of food items.
- The (Name of the Fruit) Flavoured is required to be mentioned on a food product that does not contain specified amount of fruit.
- The carbonated beverages which do not contain any fruit juice/fruit pulp shall not be label in such a way that may mislead consumers and believing them that it is a fruit product.
- Any package/container of food article shall not use the word pure.

13.8. Restriction on Advertisement

The advertisement should be in accordance with the Food Safety and Standards Act, 2006 (34 of 2006) or the rules or regulations made there under. Upon false advertisement, penalties will be applicable according to FSS Act 2006.

13.9. Food Articles Exempted from Labelling

If the surface area of the package/containers is less than 100 centimeters square, the list of ingredients, Batch number/lot number or code number, instruction for use on the label is exempted. But all these information shall be mentioned on the secondary packaging or wholesale packages.

While incase of packages/containers, where total surface area is less than 30 centimeters square the manufacturing date or expiry date or best before date, the label on the packages/containers shall be exempted. But with a condition that all these information shall be mentioned on the secondary packaging or wholesale packages.

The requirement of list of ingredients shall be exempted in case of liquid products marketed in bottles and when such bottles are intended to be reused for refilling. The bottles that are manufactured after 19th March, 2019, the list of ingredients and nutritional information shall be mentioned on the label.

Date of manufacturing is not required to be given on the label of packaged food articles in case of food with shelf-life of not more than seven days. In such case, use by date shall be mentioned on the label. The list of ingredients, nutritional information, Date of manufacture/ packing, best before, expiry date, labelling of irradiated food and vegetarian logo/non vegetarian logo, may not be specified in case of multi piece packages.

13.10. Information Regarding Deficiency, any Addition or Admixture, if Any in the Food Article

The statement on the label like "This (name of food) contains an admixture of not more than (quantity of added admixture) percent of (name of the admixture)" needs to be given.

13.11. Declaration on Trans-Fats on Food Items

Industrial trans-fats are prepared by adding hydrogen to liquid vegetable oils to make them more solid and to increase the shelf life of the food items. Trans fats are present in partially hydrogenated vegetable oils, vanaspati, margarine, bakery shortenings and can also be found in baked and fried foods.

Trans fat free logo has been launched by FSSAI for the products containing industrial trans-fat not more than 0.2 gm per 100 gm of the food. The food establishments uses trans free fats/oils, are eligible to display Trans Fat Free' logo in their outletsin compliance with Food Safety and Standards (Advertising and Claims) Regulations, 2018.The use of the trans-fat free logo is made voluntary. The image of the trans-fat free logo is depicted below:

Fig. 6: Symbol for trans free fat food products

13.12. Food Claims

Claims are statements, symbols or wording made on a product label; they can be advertising or any promotional material that gives detail about the product performance/its usage/indication/application/composition. The claims on a food product should not be misleading and to be written accurately and substantiated on the basis of scientific data and studies.

FDA has given three categories of health claims which are given as follows:

(i) **Health claims** A claim that describes a relationship between a food substance and reduced risk of a disease is known as health claim. For example, any health claim given on a food product should contain two components, components of food and which component of that particular food is associated in reducing the diseased condition.

Statements that address a role of dietary patterns in maintaining good health are considered to be dietary guidance rather than health claims.

(ii) **Nutrient content claims** describe the level of a nutrient in the product, using terms such as *free, high,* and *low*. A statement such as "only 200 mg of sodium" characterizes the level of sodium by implying that it is low. Therefore, the food would have to meet the nutritional criteria for a "low" nutrient content claim or carry a disclosure statement that it does not qualify for the claim (e.g., "not a low sodium food").

(iii) **Structure/functional claims** these claims are to be meant for conventional foods and dietary supplements as well as drugs. Statements that address a role of a specific substance in maintaining normal healthy structures or functions of the body are considered to be structure/function claims.

Structure/function claims for conventional foods focus on effects derived from nutritive value, while structure/function claims for dietary supplements may focus on non-nutritive as well as nutritive effects.

13.13. Food Safety and Standards (Advertising and Claims) Regulation 2018

FSSAI has notified claims and advertisement regulations to establish fairness in claims and advertisements of food products marketed by food companies. According to these regulations, any food business operator advertising or marketing any misleading product could attract a penalty extending upto 10 lakhs under Section 53 of the Food Safety and Standards Act 2006.

Some of the pictures highlighting false claims in food products can be seen as follows:

Fig. 13.7: (A) Picture showing false claim i.e. All natural **(B)** Picture showing false claim i.e., heart healthy when the product contains Artificial fruit and cream flavors

13.15. Bar Codes in Food Packaging

Under Food Safety and Standards (Labelling and Display) Regulations, 2020, it is mandatory to provide nutritional information in the form of Barcode/Global Trade Identification Number (GTIN).

Bar codes are an important tool used in food industries to manage supply chain and track the distribution of food products. Bar codes helps in stock management of food items and also to make sure that each product is verified at every stage. Bar codes are very useful in traceability and recall programs in food businesses.

Barcodes are symbols that can be scanned electronically using laser or camera-based systems. They carry information about the object to which it is attached in a machine readable format. Bar codes helps in reducing time as compared to manual data entry which is tedious and time time taking.

Ministry of Consumer Affairs has developed a smart consumer application that helps in giving customers the information about different labels in India. The application is very easy and simple to use. Consumers can scan the barcode and get complete detailed information about product. The customer can easily check if the product is genuine or fake.

13.15. Latest Developments in Food Labelling

(i) Date marking including date of manufacturing and date of expiry shall be given at one place.

(ii) It is mandatory to provide information related to allergens, vegetarian and non-vegetarian logo for freshly prepared and served foods.

(iii) A statutory warning shall be given on labels of alcoholic beverages such as CONSUMPTION OF ALCOHOL IS INJURIOUS TO HEALTH; BE SAFE - DON'T DRINK AND DRIVE - in English language.

(iv) Nutritional information in the form of Barcode/Global Trade Identification Number (GTIN).

(v) Calorific values (in kcal per serving and serving size) shall be displayed on the menu cards or boards in the Food Service Establishments having Central license or outlets at 10 or more locations.

(vii) Fortified foods should always carry the words "fortified with .... (name of the fortificant)". It may also carry a tag line "Sampoorna Poshan Swasth Jeevan" under the logo as depicted below:

Fig. 8: Symbol for fortified food

(vii) Under Food Safety and Standards (Organic Foods) Regulations, 2017, every package of certified organic food shall carry the logo as specified below:

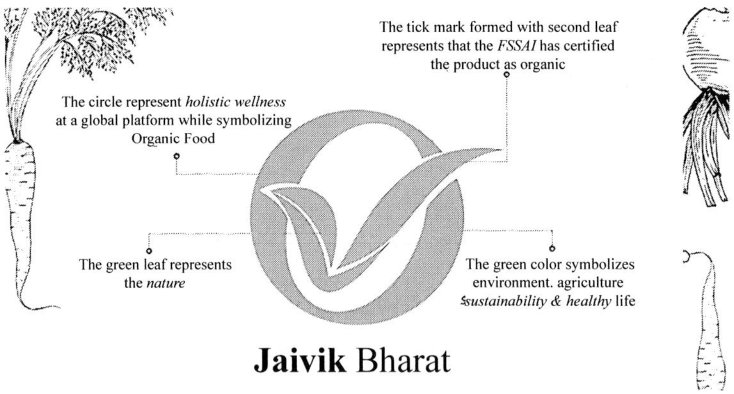

Fig. 9: Symbol for organic food

13.16. Radio Frequency Identification (RFID) Technology on the Labels

RFID is basically a wireless communication, in which person or animal, object can be specifically identified using electromagnetic coupling in the radio frequency portion of the electromagnetic spectrum. RFID is well tested technology and widely adopted by many sectors including food. these are mainly two types of RFID system i.e., Active and passive RFID system. In active RFID there is power source available with it, while in passive RFID it is not.In RFID system there are mainly three components

(i) Scanning antenna

(ii) Transceiver

(iii) Transponder

RFID interrogator constitutes scanning antenna and transciever. It can be used as permanently fixed device or portable device. It gets activated when radio wave transmit signal, after that the RFID tag sends a wave back to the antenna and gets translated into the data.

RFID tag is known as transponder. The read range of these tags depends on factors such as reader type, tag type, frequency ranges etc. The strength of power is responsible for the read range of tag, more power resulting more reading range.

RFID as comparison to bar code need not require direct line of sight and can also read several tags per unit time (seconds, minutes etc.)

References

Food Safety and Standards (Packaging and Labelling) Regulations, 2011, FSSAI

Food Safety and Standards (Advertising and Claims) Regulations, 2018, Food Safety and Standards Authority of India.

https://www.fda.gov/food/food -labelling-nutrition/label -claims-conventional-foods-and-dietary-supplements.US Food and drug administration.

Koen &Wentzel-Viljoen, 2016. Food and nutrition labelling: the past, present and the way forward. South African Journal of Clinical Nutrition. ISSN: 1607-0658 (Print) 2221-1268.

Kasbekar A., 2020. Holding true: food product claims and regulations. https://www.newfoodmagazine.com/article/130791/nutrition-and-food-claims/

Madhvapaty& DasGupta, 2015. A Study of Food Product Labelling for Products Aimed at Children. IOSR Journal of Business and Management. e-ISSN: 2278-487X, p-ISSN: 2319-7668. Volume 17, Issue 3.Ver. III (Mar. 2015), PP 88-96

McCarthy G. 2020. Barcodes on food packaging.https://www.barcoding.co.uk/barcodes-on-food-packaging/

The Standards of Weights and Measures (Packaged Commodities) Rules, 1977

14

Recycling of Packaging Material

Ankan Kheto[1], Makdud Islam[2], Ayan Sarkar[1] and Vijay Singh Sharanagat[2*]

[1]Department of Food Process Engineering, National Institute of Technology Rourkela, India
[2]Department of Food Engineering, National Institute of Food Technology Entrepreneurship and Management, Sonepat, Haryana, India

14.1. Introduction

There is a tremendous need to change the way society produces, and consumes materials to manage the environmental effect in the growing scarcity of resources. To overcome this issue, recycling process could be a great possibility that reduce the uses of natural resources, maximizethe waste utilization and produce least environmental impact. Unfortunately, packaging materials contribute 40% of solid waste in Western Europe, and 33% in US. In fact, packaging waste is likely to increase in transitioning and developing countries. Recycling process consists of collection, sorting, recovery, and management of packaging materials. As per the report of "Eurostat statistics in 2018",European Union (EU) generates 174.1 kg of packaging waste per inhabitant, Croatia contributes 67.8 kg per inhabitant and Germany produce 227.5 kg per inhabitant.

The recycling process can be defined as reprocessing of used material as a raw material or for other purposes after some physical, and chemical treatment. Basically, this process comprises of several steps, i.e., collection of used material, sorting, washing and reclamation. Collection, and sorting process is the most important steps that will decide the quality of the final product. The recyclable material should follow the standard limits as per industrial norms. Acceptance of recycled material are depended on consumer preference, socio-economic and political aspects. But market capacity/trends decide the feasibility of the recycling process.

Plastic based packaging materials is primary responsible for the major environmental and health issue. Currently annual production rate of plastic is more than 150 million tons and majority of portion is directly discarded into the environment.

In most of the cases, NIAS (Non-intentionally added substances) is often migrated from polystyrene-based packaging material to the foods and produce hippuric acid which is very toxic compounds for the human health.

14.2. Rules and Regulation

The EU body has successfully passed the law in the recycling of packaging material on December 31, 1994 by the Council Directive 94/62 and currently known as the Packaging and Packaging Waste (P&PW) Directive. Here, several issues of packaging material are addressed, and among them recycling, recovery, reuse, and prevention process are the given the higher priority to reduce the detrimental effect on environment. Moreover, other essential information has been incorporated in the revised law in 2014 by the Council Directive 2004/12/EC.

In India, several laws are implemented to encounter this issue, such as IS 2828: 2001—for plastic material; IS/ISO 14021— Environmental labels; IS 14535— guideline of recycled packaging material (plastics), respectively.

As per international/national rules and regulation, the following targets has been proposed.

- Overall recovery- 60%
- Overall recycling- 55%

Recovery target of specific materials

- Glass- 60%
- Paper/board- 60%
- Metals- 50%
- Plastic- 22.5%
- Wood- 15%

The recovery goal has been decided on the basis of recyclable materials, and proportionality of recyclable materials to meet the economically feasibility.

Apart from basic recovery target of individual materials, other critical issues also addressed into the modified law

- Lead, mercury, cadmium, and chromium VI contains in the packing materials have been restricted to 100 parts per million

- Essential standards for packaging material are revised (i.e., the composition, reusability, recoverability, and recyclability requirements) and well documented in Articles 9 and 11 of Directive 94/62/EC.

14.2.1. The Essential Requirements

Basic requirements of packaging materials (i.e., composition, reusability, and recoverable characteristics) have been suggested by numerous researchers. But the problem arises in recycling process, because at what percentage of the raw materials should be used for the process is still unknown. To overcome this issue, conformity to the European Committee of Standardization (CEN) was assigned to set the standard guidelines that would allow packaging materials to fulfil the P&PW Directive's essential requirements. In 2004, CEN released the standards for the basic requirements

14.2.2. Other Applicable Legislation

The P&PW Directive (in Article 8 and Annex I) proposed the numbering-based package identification system, which will help to identify the proper recyclable material, and assist the collection, recovery, and sorting steps. As per EU Commission Decision (97/129/EC), the packaging materials are classified into seven different categories. Also, the specifies symbols of each category has been suggested by mentoring the presence of any chemical components (Figure 1).

Fig. 1: Symbols used for the identification of substances in packaging materials

14.3. Types of Packaging Material

The consumption patterns change over the time to match the economic trend. For that, several packaging materials are used to meet the growing demand, such as plastics, papers, textiles, glass, and metals etc. Generally, paper and textile-based packaging material are flexible and light weight in nature that produce less trash. In contrast, glass and metal-based packaging materials are resistant against corrosion and stronger than others, which are normally used for high-value products. On the other hand, plastic-based packaging material captured higher market share due to the higher transparency, softness, heat seal ability, stronger strength-to-weight ratio and low cost. Moreover, plastic based packaging material are easy to prepare and customized to specific function due to the use of barrier layers, composites and additives.

14.4. Recovery and Collection Schemes

Packaging industry has paid more attention to decrease solid waste from municipal solid waste (MSW) and other household uses. To eliminate this issue, several policies and instruments have been implemented by the government, starting from city to town. Currently, most of the developed countries have reverted to extended producer responsibility (EPR) policies to reduce packaging waste, with variable degrees of trails (Table 1). Now a days, packaging waste policies only concern about waste management system. But, the proper grading of waste packaging material is not clearly addressed. Because, proper grading methods helps in better waste utilization, reusability, and recyclability. In contrast, majority of EPR programmers are focused on waste management, and cost reduction through collective collection, and recovery initiatives.

Table 1: Recovery and recycling scheme for many industrialized countries

Types	Principles	Examples
Reuse	Deposit fees	Bottles (in Germany, and Netherlands), pallets (Europe), boxes, and crates
Separate collection of recycling	Refund/deposit fees	Bottles, and cans (in California, and other states of the United States); PET bottles (Netherlands)
	Voluntary	Glass, and paper (Netherlands); PET bottles (France); plastic bottles (Belgium)
Mixed collection for recycling	Fee-based Voluntary	DSD system (Germany, and UK)

Reuse of packaging material produce environmental benefits because it does not require manufacturing and can be used directly after washing, rinsing or refurbishing in some cases. Reuse also shows several benefits because it partially offset by the expense of collection and product design requirements. In some cases, recovery of packaging materials is encouraged by refund system. For example, used bottles and crates are collected after use in many nations. Then collected bottles are transferred to the manufacturer where cleaning and refilling takes place. In Netherlands, most beer bottles have the same design, allowing most breweries to reuse them. Deposit systems has higher recovery rates (~ 90%), and reduce contamination from other containers or products. In addition, it also enhances the recycling rates. According to the study of beverage containers in Netherlands from 1984 to 1999, it was found that the reuse had the greatest impact on reducing waste, followed by material recycling. Similarly, pallet sharing policy is getting more popularity in the European transport companies for storage and transportation. Here all the participants share the wooden pallets that have uniform design and immediately repaired to reduce the environment issue. Likewise, crates and boxes are also reused in several industries and are often integrated with a deposit system.

Separate collection of reusable packaging material for recycling, as well as separate collection can ensure a greater quality of the collected material while reducing contamination from other materials. For decades, industrialized countries have used separate collection units for paper and glass, resulting in high recovery and lower recycling rates. Moreover, different classifications exist in the different collection systems. These systems can be built by refund or deposit fee structure or voluntary action. Curbside collection and drop-off systems are two types of voluntary systems where volunteer brings their trash to a specific area for the recycling process.

However, refund or deposit fee structure is the most popular method over the time to collect the recyclable packaging material in the developed countries (US, Europe). In US, some amounts of incentive (lies between 3 to 5 US cent) are paid to the customer after retuning back the used bottle/can in a particular location. As per study, it was observed that the average collection rate of recyclable packaging material was more than 93% from the last 20 years in the Michigan state (US). Likewise, the German Duales System Deutschland (DSD) and other systems charge a nominal fee to the manufacturers, based on the type of packaging material being used.

In case of voluntary systems, the recycling strategy depends on the manufacturer and consumer behavior. In this process, recyclable materials are stored/keptinbags/container in different household or industry. Then the stored materials are collected on daily basis or at regular interval for the recycling process. Sometimes, collected material are mixed to narrow down the intermediate steps form the various sources as per requirement. In Netherlands, paper and paperboard packaging is collected separately with ordinary paper, while in other countries, plastic-based packaging materials are collected in special bags for easy segregation and could produce pure and clean recycled material with higher recovery rate. Likewise, glass bottles, metal cans, and plastics are collected in a separate container to achieve the best recycling practices. Similarly, glass based-packaging materials are segregated by color (like clear, amber and green) in several countries for the effective recovery and recycling systems.

In mixed collection system, several types of packaging materials are mixed and then processed. However, processing condition of recycling material varies with the requirements and the type of packaging material produced. Paper-based material is collected individually, whereas other types of materials are collected combined. The individual components are separated in the post collection recovery units. This process reduces the transportation and recovery cost but overall quality of the final material is decreased. It offers lower recycling/recovery rate, if not planned properly but widely used in UK, and US.

14.5. Recycling Techniques

14.5.1. Recycling of PHA (polyhydroxyalkanoate) Based Polymers

Currently, the use of PHA-based material has increased significantly due to its low environmental impact, easiness to recycle and versatile applicability. Recycling process of PHA can be subdivided into three different ways, i.e., mechanical recycling process, chemical recycling process, and biological recycling process. Moreover, some of the advance separation methods are listed in Table 2 (Singh et al., 2017).

Table 2: Advance techniques for polymer separation

Separation techniques	**Principles**
Laser-induced breakdown spectroscopy (LIBS)	Spectral analysis is used for identification
Tribo electric separation	Surface charge transfer events allow for identification
X-ray fluorescence	X-rays are used as a source
FT-IR (Fourier transformed infrared technique)	The spectra of unknown samples were compared to the spectra of several model polymers.
Froth flotation method	Separation based on material density
Magnetic density separation	Based on the density differences between the materials
The hyper spectral imaging (HSI) technology	Analyze the image spectrum

(a) *Mechanical recycling*

Mechanical recycling process is suitable for achieving the homogeneous characteristic of material. Firstly, required materials are separated through screens, manual sorting process, sensor based automatic system, etc. to eliminate the toxic substance, unwanted material and so on. After that, shredding, crushing or milling operation takes place for the further uses, like packaging material. Initially, larger particles were break into smaller size (50-100 mm), and then converted into fine particles using jaw crushers, hammer mills etc. ranges from 10 to 50 μm. After that, the particles are sieved (cyclone/air zig-zag separators) and used for recycling process.

For example, extrusion and compression molding is used to prepare the recycled in PHA (polyhydroxyalkanoate) polymer.

(b) *Chemical Recycling*

Chemical recycling process involves depolymerization and refining through pyrolysis, cracking, gasification, chemolysis, enzymatic hydrolysis and hydrocracking. In depolymerization process, polymer chain breakdown into smaller low-molecular weight materials or monomers. In case of PHB (polyhydroxybutyrate) pyrolysis process, crotonic acid and oligomers are formed to form the polycrotic acid and vinyl acetate which is used to prepare the dental cement, plasticizers, herbicides and cosmetic products.

Table 3: Comparison of mechanical and chemical recycling process

Methods	**Techniques**	**Advantages**	**Challenges**
Mechanical recycling	Flotation	Cost effective technology	Limited for binary mixtures
	Melt filtration	Non-melting impurities can be removed	Fluctuation in pressure
	FT-NIR	Post-drying not required	Difficult to understand the actual compounds
	Electrostatic separation)	Efficient for various plastics	Pre-treatment is required
Chemical recycling	Chemolysis	Extremely suitable for value-added items	High cost, and large volumes are required
	Pyrolysis	Suitable for highly heterogeneous plastic compounds	Very complex chemical reactions
	Fluid catalytic cracking	Narrow product outcome	The catalyst is deactivated
	Hydrogen technologies	Different elements already commercialized	Need proper investigation
	KDV process	Oxygen and halogenated chemicals are also acceptable	Preliminary stage of research
	Gasification	Syngas is an important intermediate	Generation of toxic substances

(c) *Biological Recycling*

In case of biodegradation process, organic substances are broken down into smaller compounds and converted into minerals. Biodegradation of PHA can occur in aerobic or anaerobic condition to produce intracellular and extracellular depolymerases. Moreover, higher degradation rate of the PLA (polylactic acid)/PHA blend was found at 63°C in UV irradiation chamber, whereas 90% reduction of molecular weight was observed after 180 days. On the other hand, anaerobic digestion process converts the PHA into four different types of biodegradable polyester, i.e., polycaprolactone, PLA, PHB, and poly-butylene succinate.

14.5.2. Recycling of Paper-based Packaging Material

Various steps are involved in paper production i.e., fibers preparation, addition of additives, sheet preparation then drying and coating. Quality of recycled paper is not affected too much when it is prepared from recovered fibers but removal of impurities present in fiber residues is the main concern point.

(a) Re-pulping

Pulping is the procedure that removes the non-cellulose portion from native wood, preparing the material for paper manufacturing. It comprises of an alkaline treatment at pH 8-10, carried out in appropriate reactors (pulpers) of various dimensions and sizes, with the goal of defibrizing and obtaining a certain pulp consistency. The term 're-pulping' refers to recovered fibers and describes the process of dissolving recovered paper in water for further processing. Re-pulping is the initial phase and is related with mechanical cleaning instruments such as ink removal.

(b) Mechanical Cleaning Operation

Mechanical cleaning is achieved by washing screening and suspension, respectively. During cleaning, unwanted chemical/fabric impurity i.e., sand, metal, glass, plastic, and paint residues are removed by using different types of mechanical tools. After the pulping stage, the control of contaminants eliminated during pulping is the essential parameter that determines the efficiency of these mechanical cleaning approaches.

(c) Flotation

Flotation is the most frequent method for de-inking. The paper pulp is insufflated with airflow in the presence of surface-active chemicals. In these conditions, ‘flotation bubbles’ emerge on the surface of which inks and pigments with particle sizes of 20 to 100 μm, as well as pieces of fibers and other paper additives, are suspended. The foam that has formed on the pulper’s surface is subsequently removed. The particle size of inks containing components soluble in alkaline solutions is limited. Inks containing such components break down into particles of 1 μm or less, which are unsuitable for flotation. To eliminate this issue, the paper pulp is washed again and again, until the inks start to float. Moreover, coarse particles produced by cross-linked inks or inks coated with cross-linked varnishes, such as through UV cross-linking technology are too heavy to float. In this situation, dispersing the particles and resubmitting them to the floating process is still a possibility.

(d) Hot Dispersion

This is one type of thermal treatment which is used on high-consistency materials and involves subjecting fibers to strong mechanical stresses as well as vapor or steam heating at 60 to 100°C, depending on the method. Sometimes, the thermal treatment is combined with chemical treatments. The goal of hot dispersion is to reduce chemical and microbiological contamination, particularly in procedures that employ recycled water for cleaning and de-inking.

(d) Chemical Treatment

Chemical treatment is also known as 'bleaching'. This process consists of treatment with biocides, slimicides, whitening agents, enzymes, ozone, hydrogen peroxide, oxygen and other chemicals. The main purpose of this process is to improve brightness while reducing contamination.

(d) Restabilization Technique

Restabilization technique is a mechanical recycling technique where direct reprocessing of melting phase occurs. It is the most effective, economical, and reliable methods in recycling process. But several problems have been encountered due to the inferior quality of the recycled products compared to the virgin material. In recycling process, irreversible chemical changes occur in the polymer structure and deteriorates the physical and mechanical properties.

14.6. Application of Recycled Materials

In general, recycled packaging material is directly or indirectly gets in contact with the food material as a functional or protecting barrier. Though, recycling process has several advantages, but some quality issue still arises (Table 4).

Table 4: Uses of different recycled packaging materials

Types	Types of materials	Uses
Paper packaging	Recycled fibers	Type II (dry, non-fatty) and Type III foodstuffs (food shelled, peeled)
	Combination with plastic or aluminum	Type I (aqueous and/or fatty) like dairy products, chocolate
	Cardboard combination with plastic or aluminum	Pizza trays
Plastic packaging	PET	Production of carpets, bottles, direct food contact material.
	HDPE	Shopping bags, detergent, containers of detergent
	LDPE, LLDPE	Films

Types	Types of materials	Uses
Biobased	PLA, PHB	Packaging materials of orange juice, mushrooms, salad dressing, and yoghurt
	Starch/polyethylene films, bags, laminate pouch	Ground beef, fruits and vegetables, bread, cheese, cut vegetables, etc.
	Chitosan/ gluten film, cellulose-based box, polycaprolactone, and zein	Fresh produce, mangoes. Mushrooms, and frozen products
	PLA films, PLA-coated paper cups & lids, trays, and punnets	Packaging of fresh foods, pasta, beverages, bell peppers, and cookies Manufacturing of plates, cutlery and straws
	Starch, limestone, and fiber mixtures	Plates, bowls, food service wraps and salad containers
	Starch, and paper/pulp mixtures	Meat products, french fries and chips
	Starch films laminated on paper	Cutlery, cups and plates

(Mattsson & Sonesson, 2003; Barron et al., 2002; Barron, 2001)

For indirect contact type packaging material, it protects the food material for external threats during storage and transportation. In addition, effectiveness/ efficiency of recycling process depends on thickness, type, molecular weight, chemical structure, mobility, concentration, type of food material, storage time, filling condition and temperature.

3-layer polypropylene cups: Franz et al. (1994) conducted a study with 3-layer polypropylene (PP) cups for dairy product where 95% recycled were PP used. During this study, dairy products were stored in refrigerated condition for short time and qualitative analysis of raw & finished product and migration testing were conducted. Results were found within the range of international quality and safety standers and can be use without any risk.

Multi-layer PET bottles: Franz et al. (1996) developed multilayer PET bottle where internal layer was prepared from recycled PET. Result found, non-significant migration rate for recycled PET bottle having 186±39 μm thickness at 10 days (40°C) and 6-month (25°C) storage duration. So, multilayer PET bottle can be used for soft drinks.

Multi-layer PET films: Piringer et al. (1998) studied3-layer PET films with 20 to 60 μm thicknesses and analyzed the diffusion rate under various contact medium. They found, diffusion rate through internal layer to outer layer during co-extrusion process was up to the range and can be used as a packaging material. Franz et al. (1997) also studied three layers high impact polystyrene

(HIPS) with 100 to 200 μm thicknesses and found the results were acceptable range. Also found, HIPS layers were effective against 50% ethanol solution medium for 76 days at 40°C and for aqueous food also at refrigerate or room condition.

Recycled direct contact type material was preferable for lesser manufacturing cost. Most of the time, it uses in PET/HDPE bottle for soft drinks or fresh milk. Today, monolayer PET/HDPE bottle are used in Austria, Belgium, France, Germany, Netherlands, Norway, Sweden, and Switzerland. Sometime, oligomer is found in recycled HDPE bottle after deep cleaning. To eliminate this problem electronic nose is used in that process which detect solvents and volatile compound.

References

Barron, C., Varoquaux, P., Guilbert, S., Gontard, N., & Gouble, B. (2002). Modified atmosphere packaging of cultivated mushroom (*Agaricusbisporus* L.) with hydrophilic films. Journal of Food Science, 67(1), 251-255.

Bastioli, C. (2001). Global status of the production of biobased packaging materials. Starch-Stärke, 53(8), 351-355.

Chappin, M. M., Hekkert, M. P., & Van Duin, R. (2005). Decomposition analysis of Dutch beverage packaging waste: an analysis of material efficient innovations. Resources, conservation and recycling, 43(2), 209-229.

Franz, R., Huber, M., &Piringer, O. G. (1994). Testing and evaluation of recycled plastics for food packaging use—possible migration through a functional barrier. Food Additives & Contaminants, 11(4), 479-496.

Franz, R., Huber, M., Piringer, O. G., Damant, A. P., Jickells, S. M., & Castle, L. (1996). Study of functional barrier properties of multilayer recycled poly (ethylene terephthalate) bottles for soft drinks. Journal of Agricultural and Food Chemistry, 44(3), 892-897.

Franz, R., Huber, M., &Piringer, O. (1997). Presentation and experimental verification of a physico-mathematical model describing the migration across functional barrier layers into foodstuffs. Food Additives & Contaminants, 14(6-7), 627-640.

Mattsson, B., &Sonesson, U. (2003). Environmentally-friendly food processing (Vol. 91): Woodhead publishing.

Miranda, R., Monte, M. C., & Blanco, A. (2013). Analysis of the quality of the recovered paper from commingled collection systems. Resources, conservation and Recycling, 72, 60-66.

Notarnicola, B., &Tassielli, G. (2003). GM Nicoletti University of Foggia, Italy. Environmentally Friendly Food Processing, 306.

Piringer, O., Franz, R., Huber, M., Begley, T. H., & McNeal, T. P. (1998). Migration from food packaging containing a functional barrier: mathematical and experimental evaluation. Journal of Agricultural and Food Chemistry, 46(4), 1532-1538.

Singh, N., Hui, D., Singh, R., Ahuja, I., Feo, L., & Fraternali, F. (2017). Recycling of plastic solid waste: A state of art review and future applications. Composites Part B: Engineering, 115, 409-422.

Van Sluisveld, M. A., & Worrell, E. (2013). The paradox of packaging optimization–a characterization of packaging source reduction in the Netherlands. Resources, Conservation and Recycling, 73, 133-142.

Index

G

H

I

K

L

M

W

Y

Z